Astronomers' Observing Guides

Other Titles in This Series

John W. McAnally

JUPITER
and How to
Observe It

 Springer

John W. McAnally
USA
cpajohnm@aol.com

Series Editor
Dr. Mike Inglis BSc, MSc, PhD.
Fellow of the Royal Astronomical Society
Suffolk County Community College, New York, USA
inglism@sunysuffolk.edu

British Library Cataloguing in Publication Data
A catalogue record for this book is available from the British Library

Library of Congress Control Number: 2007932968

Astronomers' Observing Guides Series ISSN 1611-7360

ISBN: 978-1-85233-750-6   e-ISBN: 978-1-84628-727-5

Printed on acid-free paper

© Springer-Verlag London Limited 2008

9 8 7 6 5 4 3 2 1

Springer Science+Business Media
springer.com

In memory of my father and mother, John R. McAnally, Jr., and Margaret McAnally. I am eternally grateful for their love and encouragement.

And in memory of my mother-in-law, Mary Dicorte, who always asked how the book was coming along. I will miss her prayers, love, and kind ways.

And to my wife, Rose Ann, for her infinite love and patience.

Acknowledgments

As with any endeavor of this magnitude, I owe much to friends and colleagues. I wish to thank all of my fellow amateur astronomers who have so graciously contributed images for the illustrations in this book, including Donald C. Parker, Ed Grafton, P. Clay Sherrod, Eric Ng, Damian Peach, Christopher Go, Cristian Fattinnanzi, Brady Richardson, Trudy LeDoux, and Dave Eisfeldt.

I am also grateful to the editor Mike Inglis and to the publishers at Springer for the invitation to write this book, and for their patience in allowing the time I believe was necessary to make it a good one.

I greatly appreciate my colleagues, the staff of the Association of Lunar and Planetary Observers, for their support and friendship. Likewise, my friends in the Central Texas Astronomical Society have given me great moral support and encouragement.

Finally, I am deeply grateful to Amy Simon-Miller, Glenn Orton, and Scott C. Sheppard for their assistance in gathering papers and materials, and for many, many useful discussions. Their support and friendship has been invaluable to me.

Author contact information:
By e-mail: cpajohnm@aol.com
or
through the Web site of
The Association of Lunar and Planetary Observers

Contents

Section II **How to Observe the Planet Jupiter**

Introduction

Jupiter and How to Observe It

Welcome to a wonderful pastime! Observing the planets and learning something about them is an activity that anyone can do. I often liken amateur astronomy to the game of golf. Anyone can take up the sport. You can spend lots of money for equipment or you can be more frugal. You can participate at any level you wish and you can start when you are young and continue until you are old, all of your life at any age! However, amateur astronomers have one great advantage; we don't have to complain about our golf scores! My interest in astronomy began in the 1960s, not in science class but in reading class. We read a story in the eighth grade about the Hale 200-in. telescope on Mount Palomar, and how George Hale raised the money so it could be built. I am not sure what happened, but something in me just clicked and I knew that somehow I had to get into astronomy. My parents were poor, so my first telescope was inexpensive, small and hopelessly inadequate; yet, I remember going out with it every clear night. Later in high school I purchased a telescope that was still small but much better optically, and my views became much more clear. The planets especially have always fascinated me with their bright appearance and motion against the background stars. Whether observing visually or taking images through a telescope, I continue to be intrigued by what can be seen on their surfaces, by what changes and what stays the same!

In writing this book it is my hope that after reading it the beginner, who is just starting out, will acquire enough knowledge from it to be able to go to the telescope and make a meaningful observation the very first time. The methods and procedures described are not, for the most part, overwhelming or difficult; they simply require patience, care, and attention to detail. I believe the advanced amateur will also find enough here to be challenging, especially the more advanced procedures of imaging and reducing and reporting real data that is scientifically valuable. I have tried to follow a logical approach. As with any new endeavor, it is important to understand terminology and scientific notation about the subject to be studied, before the study is undertaken. Speaking the language is important and I have tried to make 'Jupiter speak' a little less daunting. It can also be helpful to have an understanding of the subject's past history and to think about what might occur in the future.

Section I of this book will discuss much of what we already know about Jupiter, and will hopefully provide a good grounding in the planet in preparation for Sect. II. Section II will discuss in some detail how to observe the planet, record data in a meaningful way, and report it. There is so much about Jupiter to know and understand. My own lifetime journey through this learning process has been a most enjoyable experience. I hope you enjoy yours. Every night can be a new adventure!

John W. McAnally
Assistant Coordinator for Transit Timings
Jupiter Section
Association of Lunar and Planetary Observers
2124 Wooded Acres
Waco, Texas 76710
cpajohnm@aol.com

Section I

Chapter 1

The Earliest Observations

1.1 Known to the Ancients

Jupiter is so bright in the night sky that it can easily be seen with the naked eye; in fact, among the planets, only Venus can shine brighter. Being so bright it was known to man long before the invention of the telescope. Ancient civilizations around the world knew of its wanderings and made attempts to predict its behavior against the stars. In mythology, Jupiter was the chief god of the Romans. The Greeks referred to Jupiter as Zeus. I can imagine that even prehistoric man would have noticed Jupiter, shining so bright against the other star-like bodies. We can think of Jupiter as our chief planet in the solar system. As we'll see, we might not even exist without this giant planet!

1.2 Galileo Galilei and Discovery of the Galilean Moons

Galileo Galilei was perhaps the first person to effectively use a telescope to explore the heavens, and is credited with being the first person to use a telescope to look at Jupiter. In January 1610 he noticed three star-like objects lined up in a row in Jupiter's equatorial plane (he eventually discovered a forth one). This alignment apparently aroused his deep curiosity and he eventually came to the conclusion that they must be in orbit around Jupiter! What a discovery! Seeing that another planet had bodies in orbit about itself, and knowing of the problems with the orbital theories of the time, Galileo came to the further conclusion that the Earth must not be the center of the motions that were observed in the universe. Having previously been encouraged in his other scientific studies by the Church in Rome, Galileo made his findings known to the Pope. Much to his disappointment, the Church soon took exception to his assertions that Earth was not the center of the universe and forbade him to continue his research or to even discuss it openly. He was subsequently placed under house arrest. Of course, we now know that Galileo was correct, but at the time Jupiter presented him with what was truly a life-threatening situation! These four moons are now known as the 'Galilean moons'.

1.3 Cassini and the Great Red Spot

After Galileo, as the quality of lenses and telescopes improved, observers began to detect markings on Jupiter's surface. In 1665, Giovanni Dominico Cassini discovered a "permanent spot" on Jupiter and followed it on and off for several years. Cassini also discovered Jupiter's equatorial current, the flattening of its poles, and its limb darkening. Later, when the Great Red Spot was recognized in 1879, it was suggested that this was a rediscovery of Cassini's spot. However, there is really no empirical evidence to support this, and we must be careful not to state this as fact [1].

As the years went by and telescopes and lenses continued to improve, more and more discoveries were made regarding Jupiter. Many of the people making these discoveries would be considered amateurs today. However, these amateurs were serious, dedicated, careful observers. And as we will see, there continues to be room in astronomy for amateurs today, and perhaps more so than ever!

1.4 In Good Company

The amateurs of today are in very good company with many notable observers who have led the way before us such as Bertrand Peek, Hargreaves, Phillips, Molesworth, and Elmer Reese; and with contemporaries of today such as Miyazaki, Don Parker, Phillip Budine, John Rogers, Walter Haas, Olivarez, and many, many others. If not for these observers, the visual record of Jupiter would be sparse indeed. We can take pride in helping to continue the works of so many wonderful amateurs.

The telescopes and related equipment are better than ever. I can only imagine what Bertrand Peek would have given for a good web cam or CCD camera! How much easier our work is compared to theirs; yet, we can only hope to measure up to their discipline, their persistence, and their attention to detail!

So, from Galileo to today many years have gone by and yet, Jupiter still begs to be observed! What will we discover next year, and the next, and the next?

Chapter 2

Jupiter's Place in the Solar System

Whether our study of Jupiter is casual or serious, it will be helpful to understand some basic facts about the planet, including simple nomenclature. This knowledge will help us in our own study, and it will allow us to understand what others say and write about the planet. As I have found over the years, there is always something new and exciting to learn, something new to be discovered and revealed, but we must understand the language.

So, where does Jupiter stand in the scheme of things? Our solar system is comprised of many bodies, large and small. We were taught about the nine planets in school, and their order in distance from the Sun. Recently, the International Astronomical Union changed the classification of Pluto, and it is no longer officially classified as a planet in the simple sense. Now we have eight planets and all manner of other bodies.

2.1 Physical Characteristics

Jupiter displays a series of bright zones and darker belts, generally running parallel to its equator. Figure 2.1 illustrates the globe of Jupiter with the belts and zones that are usually visible. Not all features will be visible at all times, as belts and zones are prone to brighten, darken, become larger or smaller, or even disappear from time to time.

Jupiter is the fifth planet from the Sun. It is a gas giant, having no surface as we think of on Earth. Its volume is so large, that if it were a hollow sphere, all the other planets would fit easily inside with room to spare. Even mighty Saturn is only about one-third its mass. However, Jupiter's density is so low that if there were a water ocean large enough, Jupiter would float on its surface!

Jupiter's large mass is of extreme importance to the solar system and especially to Earth. Jupiter's mass perturbs the orbit of nearly every planet in our solar system. It also influences the orbits of smaller bodies that come into the inner solar system from the Kuiper Belt and the Ort Cloud. Jupiter's mass and strong gravitational influence has a tendency to either sweep up small bodies that cross its orbit, or to eject them from the solar system entirely. This solar system 'vacuum cleaner' made it possible for Earth to survive long enough for life to form and evolve. Without this protection, the bombardment of Earth would occur too frequently by bodies

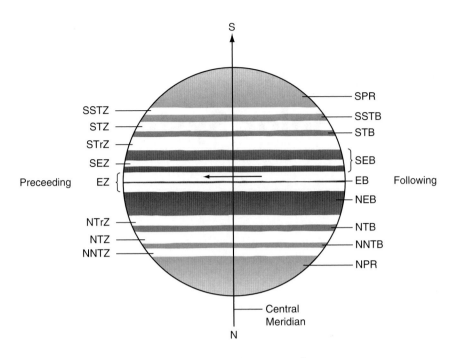

Fig. 2.1. The belts and zones of Jupiter (Credit: John W. McAnally).

Table 2.1. Jupiter's physical and orbital characteristics compared to Earth		
	Jupiter	Earth
Equatorial diameter (km)	143,082	12,756
Polar diameter (km)	133,792	12,714
Rotation periods		
		23h 56m 4s
System I	9h 50m 30.003s (877.90°/day)	
System II	9h 55m 40.632s (870.27°/day)	
System III	9h 55m 29.711s	
Axial tilt	3.12°	23.44°
Mass	1.899×10^{27} kg	5.974×10^{27} kg
Density	1.32g cm^{-3}	5.52g cm^{-3}
Surface gravity	2.69g	1.00g
Mean distance from Sun	5.20280 AU	1.00000 AU
Orbital eccentricity	0.04849	0.01671
Period (sidereal)	4,332.59 days	365.26 days
1 astronomical unit (AU) = 149,597,870 km [2]		

too large for Earth to survive as we see it today. The recent 1994 collision of Comet Shoemaker-Levy 9 with Jupiter is a great example of Jupiter as protector of the solar system.

Jupiter exhibits differential rotation; that is, different latitudes of the planet have different rotation rates. Generally, System I includes the latitudes from the north edge of the south equatorial belt, all of the equatorial zone, to the south edge of the north equatorial belt. System I also includes the south edge of the north temperate belt. System II includes the rest of the planet. Since amateurs in the past have observed Jupiter in visible wavelengths, it has been common practice for them to refer to System I and II. Professional astronomers have generally used a third rotation system, System III. The System III rotation rate is related to a radio source on Jupiter that rotates with the planet at a specific rate. Since these three rotation rates are different, we must designate which system we are referring to when we speak of longitudinal positions on Jupiter. Depending upon the latitude at which a feature appears on Jupiter, amateurs refer to System I or II longitude. This usage will become more apparent in the section of this book dealing with transit timings. Table 2.1 summarizes Jupiter's physical data and orbital characteristics.

2.2 A System of Basic Terminology and Nomenclature

Like most sciences, planetary astronomy comprises a language of special terms and nomenclature. Understanding those associated with Jupiter will facilitate our discussions and explanations, since this scientific shorthand can actually help to keep our discussions simple and unambiguous. Years ago, A.L.P.O. Jupiter Section Coordinator Phil Budine suggested a simple, straightforward system that we can still use today. There are abbreviations for the terms and nomenclature of dark and bright features, and for the belts and zones; so, some of the more common terms and abbreviations are shown in Tables 2.2 and 2.3. Various dark and bright features can be seen in the belts and zones at any given time. Some of the features most often seen are illustrated in Table 2.4. These illustrations are modeled after illustrations used by past A.L.P.O. Jupiter Recorder Phillip Budine.

A simple example can help us understand how we put this terminology into use. Figure 2.2 shows a large condensation, or barge, on the north edge of the north equatorial belt. This feature would be described as, 'Dc L cond N edge NEB'; which literally means 'dark center, large condensation, north edge, north equatorial belt.' So, you see how in simple, straightforward notation we have completely described the feature and where it resides. If we were describing a bright feature we would use the designation 'W', instead of 'D'. Later, when we discuss central meridian transit timings, you will see how we combine this description with the longitudinal position of the feature to turn this kind of observation into real, meaningful data.

As we will see in Sect. II of this book, your observations are only valuable if they are properly recorded and notated. The system of nomenclature presented here should allow anyone to accomplish this task. Organizations such as The Association of Lunar and Planetary Observers (A.L.P.O.) and the British Astronomical Association (BAA) have standard observing forms that the observer can use to record observations. Many other organizations around the world also have standardized forms. Standardized observations greatly facilitate the gathering

Table 2.2. The basic nomenclature and abbreviations for Jupiter's belts and zones

SPR	South Polar Region
SSTB	South South Temperate Belt
STZ	South Temperate Zone
STB	South Temperate Belt
STrZ	South Tropical Zone
SEB	South Equatorial Belt
SEZ	South Equatorial Zone
EZ	Equatorial Zone
EB	Equatorial Band
NEB	North Equatorial Belt
NTrZ	No rth Tropical Zone
NTB	North Temperate Belt
NTZ	North Temperate Zone
NNTB	North North Temperate Belt
NPR	North Polar Region

Table 2.3. Basic nomenclature used for transit timing observations

Dark marking	D
White or bright marking	W
Center	C
Preceding	P
Following	F
North	N
South	S
Large	L
Small	Sm
Projection	Proj
Condensation	Cond
Central Meridian	CM
System I	(I) or CM1 or L1
System II	(II) or CM2 or L2
System III	(III) or CM3 or L3

and recording of data and its subsequent use by the professional community and other amateurs.

I strongly encourage you to use this standard system of notation. Not only will its use make you a better planetary astronomer, it can even add some anticipation and excitement to your endeavors. What features are you going to be able to record tonight? Will they be the same tomorrow night, or next week? I think you will find it fascinating to learn that some features are long-lived and some are not! You are going to learn so much about Jupiter, and you will be amazed at how easily you retain what you have learned when you observe in this fashion!

Table 2.4. Basic nomenclature for dark and bright features commonly seen on Jupiter (Credit: John W. McAnally)

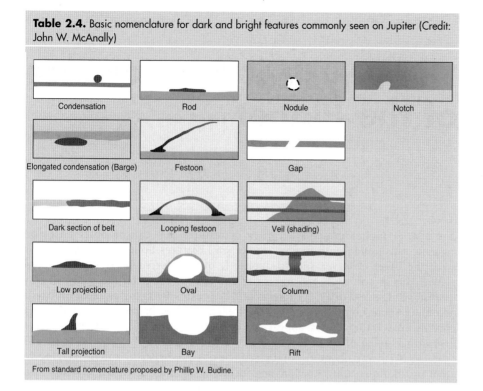

From standard nomenclature proposed by Phillip W. Budine.

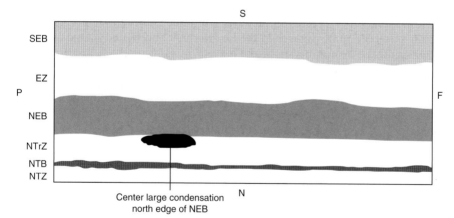

Center large condensation
north edge of NEB

Fig. 2.2. Example of a large condensation depicted on the northern edge of Jupiter's North Equatorial Belt (Credit: John W. McAnally).

Chapter 3

The Physical Appearance of the Planet

The disk of Jupiter presents a variety of features that can be observed by an amateur astronomer with modest equipment. Features both obvious and subtle await the eager observer. Indeed, Jupiter is often referred to as "the amateurs' planet", due in part to its enormous size and angular diameter that makes it easy to observe. The use of CCD cameras and web cams by many amateurs today is becoming more commonplace, with excellent images being obtained showing great detail. It is the physical appearance of the planet and ease of observation that first attracts most of us. In this chapter we will deal with the physical structure, characteristics, and phenomenon that can be observed visually and by imaging; in other words, those things that anyone, including an amateur with a telescope, can see. Specifically, we will examine non-vertical structure in Jupiter's clouds, winds, jet streams, and color.

The observation of features on the surface of Jupiter, or rather in its cloud tops, and the observation of changes in the longitudinal and latitudinal positions of features, and the determination of the movement of features with regard to the speed of various currents and the appearance and reappearance of other phenomena, has been a mainstay of amateur observations since the beginning of recorded observations more than 150 years ago. While there is much to Jupiter amateurs cannot observe directly, these things we can see. Amateurs today continue this wonderful tradition of observational astronomy.

3.1   Common Visual Markings

Although we should never approach the observation of Jupiter with any preconceived notion of what may be seen, it can be useful to have an understanding of the type of features that may be present at any given time. The bright ovals, eddies, and condensations are evidence of great turmoil and chaos in the visible atmosphere of Jupiter. Features seen in Jupiter's cloud tops are not stationary, and may be short or long lived. Knowing this, I think the observation of these features is all the more tantalizing. You may wish to review the figures and tables in Chap. 2 again while studying this chapter.

Generally, the planet is made up of clouds that have organized themselves into dark belts and bright zones, due mainly to the rapid rotation of the planet. The belts and zones are defined by powerful jet streams that are permanent winds blowing eastward or westward. Dark spots and bright ovals are storms that drift eastward or westward at slower rates. The rapid rotation of the planet causes all motions to be channeled along lines of latitude; thus the belts and zones run east and west. See Fig. 2.1 in Chap 2, which depicts the placement of belts and zones and nomenclature.

3.1.1 The Polar Regions

To the visual observer, the polar-regions most often present a gray, dusky appearance, absent specific features. Indeed, during the apparitions of 2000–2001 and 2001–2002, most visual observers reported no features at all. A few observers using large instruments with exceptional image contrast reported an occasional small, bright oval near the latitude of the NNTB or SSTB. However, these sightings were never widely observed and thus, difficult to confirm. Occasionally there is an exception to the inactivity usually noted in the polar-regions, as reported during October 1997. During that apparition, several independent observers observed faint dusky markings, slightly more intense than the surrounding region. Among those observers, I was able to observe one feature well enough to make one set of transit timings. This occurred on the night of October 26, 1997 U.T. (Fig. 3.1). Quoting from my own observing log, "The NPR is also exceedingly

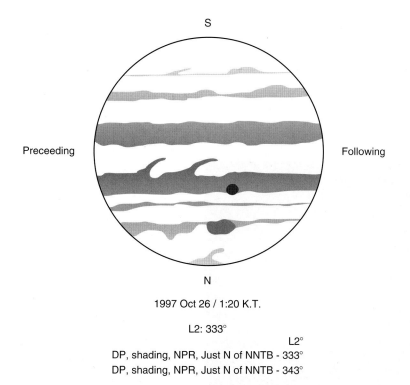

1997 Oct 26 / 1:20 K.T.

L2: 333°

                                               L2°

DP, shading, NPR, Just N of NNTB - 333°

DP, shading, NPR, Just N of NNTB - 343°

Fig. 3.1. A full disk drawing of Jupiter showing the shading or veil in the north polar region of Jupiter on October 26, 1997 (Credit: John W. McAnally).

faint. A dusky patch or mottling is seen by me for the first time at the southern edge of the NPR (these had previously [recently] been reported by others). This mottling was very distinct and unmistakable! It was fairly extended in longitude. It was seen clearly enough that I was able to make a CM transit timing of the preceding and following edge. This may be the first measurement (CM transit timing of such a feature, at least during this apparition) reported to A.L.P.O. (this observation was made in integrated light). The mottling was also seen in green (W56) light. Not seen in yellow (W12) light." Alas, the feature disappeared and was not observed again. However, their existence was confirmed beyond doubt when these dusky markings were captured in a few CCD images, coinciding with the longitudinal position observed by me and several other observers. These dusky markings were first seen visually and later also confirmed by CCD imaging. A fine example that visual observations are still of value.

Often, the polar-regions seem to spread their gray appearance all the way to the NTZ or the STZ. Occasionally, however, subtle, lighter zones can be made out, divided by thin, grayish belts barely seen. Observing the often too subtle NNNTB and NNTB, their southern counterparts, and intervening zones, is extremely difficult. Consequently it is very difficult to collect current, useful data on the drift rates and wind currents in these regions. Occasions have been rare in which any useful transit timings have been obtained from the polar-regions. However, when bright or dark features present themselves in these belts and zones for long enough, the observations that can be obtained are of great importance. As recent as July 2006, CCD and webcam images continued to reveal a north polar region that was mostly unremarkable (Figs. 3.2 and 3.3). However, the improved camera technology consistently revealed light and dark dusky markings in the region. With advances in CCD and webcam technology, we can expect amateurs to obtain images with such high resolution that features in the polar regions will be captured consistently with enough resolution that measurements of currents and drift rates in this region will not only become possible but routine.

05:19 UT

CM1=98.5 CM2=136.8 CM3=57.8

Fig. 3.2. A webcam image of Jupiter demonstrating the lack of detail in Jupiter's north polar region on February 11, 2004. South is to the top right of image. (Credit: Donald C. Parker).

Fig. 3.3. A webcam image showing a little more activity in Jupiter's north polar region, although the region is still unremarkable, on April 12, 2006. South is to the top right of image. (Credit: Donald C. Parker).

3.1.2 The North–North Temperate Region

The North–North Temperate Region generally extends from 57° North to 35° North latitude. As mentioned, the North–North Temperate belt (NNTB) can be difficult to observe and is not always present (Fig. 3.4). However, when the belt is present, it is important to note its width and intensity. There may also be seen the occasional projection on its southern edge. Although these are rare, if present, careful observations and transit timings can yield valuable data concerning drift rates and wind currents. During the 2000–2001 apparition, a remarkable, dark segment appeared in the NNTB (Fig. 3.5). In fact, a disk drawing of Jupiter on August 2, 2000 shows this dark segment quite well. Similarly, CCD images by David Moore, I. Ikemura, Peach, Maurizio Di Sciullo, Antonio Cidadao, and others around the world also captured this dark segment. Observers continued to detect this segment in CCD images in early 2002, although the segment was fading and not near the intensity of 2000–2001. By April 2002, images by Don Parker revealed a NNTB segment that was broken and spread out. Thus, the feature was long-lived and yielded very useful data concerning the drift rate and wind speeds in this belt. It measured 10–15° in longitudinal length when first seen. Features such as this are very rare in the NNTB. This one presented a wonderful opportunity to check recent data about currents against that collected in prior years with Earth based telescopes and against wind speed data collected during the Voyager series. Often, the NNTB goes wanting for such data! This feature was extensively imaged by amateurs using CCD cameras but was also widely observed visually. The feature lasted for over 14 months. The NNTB was unremarkable from June 2002 to May 2004 (Fig. 3.6). However, in April 2006 imagers detected a bright oval and a little red spot (LRS) in the NNTZ. Portions of the NNTB were also seen, reddish-brown in color.

Fig. 3.4. An image of Jupiter showing a fairly dark NTB, and a discontinuous NNTB on September 30, 1998. Note the narrow NEB and bright Mid-SEB Outbreak running down the middle of the SEB. South is up. (Credit: Donald C. Parker).

Fig. 3.5. Jupiter on August 30, 2000. Note the dark segment of the NNTB and that the remainder of the NNTB is discontinuous or absent. South is up. (Credit: Donald C. Parker).

3.1.3 The North Temperate Region

The North Temperate Region extends from latitude 35° North to 23° North. While the polar-regions are often disappointing, the North Temperate Belt (NTB) contains a class of feature all its own. To the casual observer, the NTB usually presents the appearance of an otherwise featureless, thin, grayish or reddish-gray belt slightly more intense than the neighboring polar region. However, closer scrutiny reveals otherwise. The belt can be continuous, fragmented, or completely absent. During the 2000–2001 and 2001–2002 apparitions, the NTB was reported by most observers to have a reddish-brown coloration mixed in with the gray tone (Fig. 3.7). I have

04:44 UT

CM1=278.7 CM2=324.7 CM3=246.5

Fig. 3.6. Jupiter on February 10, 2004. During the 2004 apparition both the NTB and NNTB disappeared from Jupiter's disk. Notice the prominent bluish-gray features on the NEBn and the faint bluish band in the NTrZ. South is up. (Credit: Donald C. Parker).

Ed Grafton
05:12 UT
L1 133.2, L2 098.6, L3 068.5
Celestron 14-inch f/10 SCT

Fig. 3.7. Jupiter showing a moderately intense GRS and NTB. CCD image taken November 28, 2000. South is up. (Credit: Ed Grafton).

observed on more than one occasion that the coloration and intensity of some segments of the belt resemble the appearance of the north equatorial belt. The belt is also often observed to be inconsistent in its width. In the most recent apparitions, some observers have noted that segments of the belt are often split, or double, and that its intensity can vary greatly.

During the 1999 apparition, the NTB was prominent and continuous. However, the belt can also be absent. By 2001, some segments of the NTB had become rather faint. Indeed, by December 2001, images taken by Ed Grafton revealed segments of the NTB that were mostly gray in coloration and as light in intensity as the north polar region. By January and February 2002, images by Ed Grafton, Maurizio Di Sciulla, and Donald Parker revealed that, while some segments of the

NTB were the normal, reddish brown color, much of it had taken on this fainter, gray appearance. By December 2002, images by Parker revealed that the NTB had mostly vanished with only a few remaining dark segments scattered around the planet. On February 22, 2003 a CCD image taken by Eric Ng of Hong Kong, revealed a short segment of NTB, 20° long, near System 2 longitude 320°. By May 3, 2003 a CCD image by Christophe Pellier revealed a very short, faint segment of the NTB near System 2 longitude 140° (Fig. 3.8). I saw no remnants of the NTB in CCD images after that date. The NTB continues to be absent through July 2006. This gives Jupiter the appearance of having a bright alabaster zone stretching from the north edge of the NEB to the NPR with only a faint, bluish gray north tropical band south of the NTB's normal position. In some high-resolution CCD images, a faint bluish-gray band can also be made out in the north temperate zone (NTZ). Do not mistake these bands for remnants of these two belts, as they are not (Fig. 3.6). According to Peek [3] and Rogers, NTB fadings normally last for periods from 8–13 years [4]. As recent as February 2007, the NTB was absent. However, that began to change in late March and by April 27, 2007, the NTB was almost completely restored. We must be always alert because sometimes events really do happen just this quickly!

Some of the most interesting, visual features of the NTB may be the so-called rapid moving spots (RMS) that are sometimes present. According to Peek, "in 1880 there appeared for the first time in Jupiter's recorded history an outbreak of dark spots at the south edge of the NTB that displayed the shortest rotation periods that had ever been observed on Jupiter" [5]. The drift rates of the RMS are incredibly fast. These spots are seen on the southern edge of the NTB (NTBs) when present and may reside within the NTBs prograding jet stream. This jet stream is the fastest jet stream on the planet. The RMS may also be associated with the NTBs jet stream outbreaks. These outbreaks appear to have had periods of 10 years in the past and more recently 5 years [6]. Although a few suspicious features were observed during the 2000–2001 and 2001–2002 apparitions, the last notable outbreak of RMS seen visually seems to have occurred in 1997. In that year, several spots were recorded and observed throughout the apparition. These 1997 RMS had a drift rate of −56 to −57° per 30 days. These spots appear to have been associated with the North Temperate Current C (NTC C). NTB outbreaks may also result in the appearance of white spots or ovals.

Fig. 3.8. A highly detailed webcam image of Jupiter taken on January 5, 2003. Note the discontinuous, fragmented NTB and narrow NEB. The EZ appears to be very disturbed and south temperate oval BA is seen on the STBs. South is up. (Credit: Ed Grafton).

In addition to observations of the RMS, the NTB can also fade, and it can exhibit a shift in latitude. Dark spots and streaks can also appear on the north edge (NTBn) of the NTB from time to time [7]. During the 2000–2001 and 2001–2002 apparitions, several dark condensations could be observed on the NTBn. A few observers even detected faint gray festoons streaming away from some of these dark condensations, stretching into the brighter NTZ. Occasionally, a bright oval could be seen on the NTBn, intruding noticeably into the NTB. These features were easily seen in the better amateur CCD images.

Separating the NTB and the NNTB is the NTZ. The NTZ can exhibit some interesting features worth monitoring by amateurs. On many occasions the NTZ simply appears dusky gray, approaching the bland appearance of the polar-regions with no distinctive markings. At other times it can appear much brighter, being a distinct divide between the NTB and NNTB. During the 2000–2001 apparition, the NTZ was described as almost alabaster by many observers. Indeed, the zone rivaled the north tropical zone in brightness, appearing brighter than the north tropical zone to many observers. During the apparition of 2001–2002, several segments of the NTZ were observed to be rather dusky in appearance while the remainder of the zone was bright. These dusky segments may be caused by the absence of bright, high altitude clouds at these longitudes. Changes such as these in the appearance of a zone are worth careful monitoring and should certainly be reported. During 2006, the NTZ was very bright with no obvious activity.

3.1.4 North Tropical Region

The North Tropical Region generally extends from 23° North to 9° North latitude. The north tropical region of Jupiter displays some of the most active and vivid features of the planet. In my experience, more amateur observations are made of this region than any other, rivaled only by the region of the Great Red Spot (GRS).

The north tropical zone (NTrZ) lies between the NTB and the north equatorial belt (NEB). This zone usually exhibits a bright, alabaster appearance, very similar in brightness to the usual bright appearance of the equatorial zone (EZ.) This was the case during the 2000–2001 and 2001–2002 apparitions. However, during 2003 the NTrZ was brighter than the EZ, due to the coloration event in the EZ. By March 2004, disturbances in the EZ continued to make the EZ dimmer than the NTrZ. Often, low projections can be seen on the southern edge of the NTB (NTBs) extending into the NTrZ. Several dark projections were seen during the 2000–2001 and 2001–2002 apparitions and were followed reliably for several months, yielding good drift rate data. During 2006, the NTrZ was again one of the brightest zones on the planet, brighter than the EZ and absent any notable features.

The North Tropical Current (NTC) controls all the visible features in the NTrZ and the northern edge of the North Equatorial Belt (NEBn). Although there are variations in the speed of the current, it applies to all major spots whether bright or dark [8].

Sometimes a thin bluish band can be seen running through the NTrZ longitudi-nally. This North Tropical Band is not easily seen visually, but was present during the 2002–2003 and 2003–2004 apparitions as revealed by CCD images (Fig. 3.6). Even in the CCD images, the band was very subtle and not visible the entire apparition. This band has come and gone several times during Jupiter's recent history. It was evident during 2006, spanning the circumference of the planet.

On a CCD image taken by Don Parker on 2004 July 14, the NTrZ was one of the brightest zones on the planet, even out shining the EZ which had a dusky, yellow-ochre appearance over most of its width.

The North Equatorial Belt (NEB) is home to some of the most obvious and reliable features on the planet. During most apparitions, condensations, barges, festoons, ovals, and projections can be seen in the belt or on its edges. While many of the belts and zones on Jupiter are low in intensity, the NEB is often one of the three darkest features of the planet. During the apparitions of 2000–2001 and 2001–2002 the belt generally appeared solid and wide with a reddish brown coloration reported by most observers (Fig. 3.9).

Dark condensations of oval shape, large and small, can often be seen on the northern edge of the belt. The larger condensations are referred to as barges and seem to be most conspicuous when the NEBn has receded slightly, as though leaving these condensations in its place. The condensations are most often described as reddish-brown in color, a darker version of the NEB. Barges were very prominent during the 1997 and 1998 apparitions, with several being seen at different longitudes around the planet. During the 1999 apparition, the northern edge of the NEB receded at several locations around the planet, giving the NEB a very narrow appearance at these locations. This further gave the NEBn the appearance of being undulating and wavy, or of having bright bays protruding into the NEBn from the NTrZ. Dark condensations were noted at several places around the planet on the NEBn (Fig. 3.10).

During the 2000–2001 and 2001–2002 apparitions, the barges were much less conspicuous, with smaller condensations in their place. Even the smaller condensations were not so prominent during 2000–2001. However, during 2001–2002, something unusual occurred. During that apparition several barges, similar to those often seen on the NEBn, were discovered in the middle of the NEB at several positions around the planet (Fig. 3.11)! Several of these barges were prominent enough to be seen visually throughout the apparition. During one recent apparition, two of these mid-NEB condensations were reliably observed by CCD imaging. Over time, they drifted together and merged. I had the privilege of observing several of these mid-NEB barges on a night of good seeing with the 36-inch telescope at McDonald Observatory in Texas during the fall of 2001. The rich, reddish-brown

05:31 UT

CM1=83.7 CM2=277.9 CM3=239.8

Fig. 3.9. A CCD image of Jupiter, taken on October 29, 2000 revealing a wealth of detail. Surviving south temperate oval BA is seen on the STBs followed by darker material. The SEB is split by a bright SEZ. The EZ is undergoing a coloration event, with only the south most portion remaining bright. The NEB is broad, and the NTB is also very intense, both with a strong reddish-brown color. Notice the long, bright rift running through the NEB. A small, rare bright oval can be seen in the NPR. South is up. (Credit: Donald C. Parker).

04:02 UT

CM1=343.9 CM2=217.4 CM3=337.1

Fig. 3.10. Jupiter on September 30, 1998. Note how narrow the NEB was in 1998 compared to 2000 (previous image). Also note the mid-SEB Outbreak in the SEB. South is up. (Credit: Donald C. Parker).

Fig. 3.11. A CCD image of Jupiter taken on January1, 2001. Note the two dark condensations or barges imbedded in the middle of the NEB, with a bright oval between them. Most often we see these condensations on the northern edge of the NEB. Also note the split SEB with very little coloration. South is up. (Credit: P. Clay Sherrod).

coloration of the barges was beautiful to behold! The dark condensations or spots are cyclonic as observed by spacecraft [9].

By 2002 and continuing into 2003, the NEB had returned to its more normal broad appearance with an even NEBn. The dark condensations, or barges, seen in the late 1990s were for the most part absent during 2002. In place of barges along the NEBn, during the 2001–2002 apparition, at least four small white ovals were seen at the edge of, or just barely imbedded in, the northern edge of the NEB at various locations around the planet. By March 2004, these ovals had disappeared. Instead, the NEB was again narrowing at several positions around the planet. By March 2004, neither white ovals nor dark condensations, nor barges, were seen on the NEBn, except for the long-lived white spot "Z", seen on the edge of the NEBn. The NTrZ was so bright, that white spot "Z" was exceedingly difficult to make out visually, and even difficult to distinguish in many CCD images. The NEBn was very uneven and undulating. In many places, it presented a ragged appearance on the northern edge of the belt, with the reddish-brown dark material of the belt broken up by the bright, white clouds of the NTrZ.

White ovals of the NEB are anticyclonic [10] and can sometimes be seen on or near the NEBn. When the northern edge of the NEB is receded, the ovals may appear completely in the NTrZ. At other times the white ovals can be seen to protrude into the northern edge of the NEB, forming a bay or large notch in it. Occasionally, a white oval may be seen completely within the northern edge of the NEB, perfectly surrounded by the coloration of the NEB itself. This gives a striking appearance and is often referred to as a "porthole." The long-lived white oval, designated oval "Z," has been observed for some time on the NEBn. This oval was widely observed and reported during the 2001–2002 apparition. Again, features as prominent as this allow long-term observations and provide a good means for determining drift rates at the latitude occupied. Four small, but very bright and prominent ovals were also seen during the 2001–2002 apparition embedded in the NEBn at different locations around the planet. These smaller ovals were prominently absent during the five previous apparitions. Their appearance was a very interesting addition to the belt. Although none of these small ovals were seen or reported visually, they were followed consistently in CCD images for several months of the apparition. The small ovals have now been seen with some frequency in following apparitions now that CCD and webcams are producing images with ever improving resolution.

Although during many apparitions we become used to seeing the NEB as somewhat wide and solid, bright rifts can occasionally be seen running through the belt. Such was the case in 2000–2001 (Figs. 3.5 and 3.9) and even more so during the 2001–2002 apparition. During these two apparitions, many bright rifts were detected on CCD images. Several of these rifts were prominent enough to be seen visually, and I observed several with an 8-inch reflector. The rifts usually extended for some distance around the planet, with lengths of 20–30° not uncommon. Some of the rifts were so visually stunning as to give the NEB the appearance of being split into a north and south component along some segments of the belt. Rifts generally are white spots or streaks with varying shapes near the middle of the belt. They are usually oriented from south-preceding (Sp) to north-following (Nf), sheared as it were by a velocity gradient. During 2004, bright rifts were again prominent running through the NEB (Fig. 3.12). During March of that year, rifts were so prominent in a segment of the NEB as to make the belt appear faded when viewed visually through an eyepiece. Bright rifts can change in shape and length over short periods of time.

02:22 UT

CM1=183.6 CM2=39.7 CM3=327.1

Fig. 3.12. Jupiter on March 6, 2004. This webcam image reveals a bright rift running through the middle of the NEB for quite some distance. Note the low contrast of the GRS on the following limb compared to the SEB. Four small ovals are prominent in the SSTZ in this image. The NTB and NNTB are still absent. South is up. (Credit: Donald C. Parker).

Although the NEB does not go through distinct jet stream outbreaks like other belts, it can exhibit variations in its activities in addition to those already cited [11]. Although the color of the belt is generally always reddish-brown to grayish-brown the northern edges have been known to occasionally appear yellowish. The NEB can undergo latitude shifts. Over recorded history, while the southern edge of the belt has been fairly stable, the north edge has varied greatly, especially during the late 1800s to early 1900s. During the 2000 through 2002 apparitions, some segments of the NEB became rather narrow, with the northern edge retreating significantly. And, while the NEB is normally not subject to fading as the SEB can be, it has gone through fadings and revivals (restoration of the belt) in the mid-19th century and early-20th century [12].

During 2006, the NEB was again broad, and was about as dark as the SEB. The northern edge of the NEB was relatively smooth, without large undulations or bays. Long-lived white oval Z was well seen. There were also other smaller ovals, similar to the ones often found in the SSTB, seen embedded in the northern edge of the NEB. These small ovals were difficult to see visually, but were quite prominent in good CCD/webcam images. To me the color of the NEB was a distinct reddish-brown. There were a few darker reddish-brown barges seen in the middle of the NEB at various locations around the planet. There were also distinct bright rifts running longitudinally through the NEB, with a prominent one centered near 89° System II, on an image taken by Donald Parker on 15 July, 2006. This bright rift was at least 50° in longitudinal length and could be seen visually. The bright rift ran through the NEB and its preceding end opened into the EZ at 307° System I (Fig. 3.13). The rift, residing near the southern edge of the NEB, caused the southern edge of the NEB to appear uneven. There were many grayish-blue projections on

Fig. 3.13. A webcam image of Jupiter on July 16, 2006. Note once again a bright rift running through the NEB. A distinctly reddened south temperate oval BA is in conjunction with the GRS. The EZ is very disturbed and the festoons on the NEBs are very prominent. The NTB and NNTB have still not returned. South is up. (Credit: Donald C. Parker).

the southern edge of the NEB during 2006, and these projections were quite prominent. Unlike some apparitions, these projections often formed beautiful festoons trailing off into the EZ. The bases of most of these festoons were broad and dark, easily seen visually.

3.1.5 Equatorial Region

The Equatorial Region extends from 9° north to 9° south. Another prominent feature almost always seen with the NEB are the dark, bluish-gray projections on the southern edge of the NEB (NEBs), often with bluish-gray festoons trailing away from the projection into the Equatorial Zone. I have found that, even when the seeing is much less than perfect, most amateurs can observe and record these projections and festoons on the southern edge of the NEB (NEBs) with some reliability. Along with the barges of the NEBn and the GRS, these projections are among the most consistently observed features on the planet. The bluish-gray projections seen on the southern edge of the NEB move with the North Equatorial Current.

It seems during every apparition, several of these NEBs bluish-gray projections and festoons can be observed around the planet. Peek noted during one apparition that it was almost certain that, during a single hour of observation, at least one dark projection from the southern edge of the NEB would be recorded passing Jupiter's central meridian as the planet rotated on its axis [13]. Today, I still find this to be true on almost any night of observation. The dark projections are generally features that catch the eye immediately. According to Peek, they take many forms, from tiny humps or short spikes, to large elongated masses or streaks. The humps and spikes are often the points of departure of gray wisps or festoons, some of them most delicate and some quite conspicuous, that seem to issue forth from the south edge of the NEB and look as if they were dispersing like smoke in the Equatorial Zone [14]. I have had the pleasure of noting this wonderful appearance for myself on many occasions. During one observation that particularly stands out, Jupiter was at opposition and at its closest to Earth, with the seeing being near perfect. On that night, not only could several festoons be easily made out; but, so much detail could be seen inside the festoons, they appeared to have been braided! Most often, the seeing is not that good. During the apparitions of 1997, 1998, and 1999, festoons were very prominent and easily seen visually (Fig. 3.14). However, by 2002, the festoons had become thin and faint, and much more difficult to make out. Even their appearance in CCD images was remarkably unspectacular! By 2004, the festoons were becoming darker again and not quite so thin; thus, much easier to make out visually. Some of the bluish-gray projections, while not trailing prominently into the EZ, did display large, intense bases. That is to say, you could see a large, long bluish-gray feature lying on the southern edge of the NEB. This often gave the projection the appearance of a bluish-gray plateau, some nearly the length of the GRS itself, and quite easy to see (Fig. 3.15). Peek described this effect as "dark masses and streaks strikingly conspicuous and rectangular in outline" [15]. To an inexperienced or infrequent observer, these changes might not be so apparent.

To this day, no one ever described these bluish-gray NEBs features as poetically as Bertrand. M. Peek. Peek wrote, "The dark projections are generally features that catch the eye immediately. They take many forms, from tiny humps or short spikes to large elongated masses or streaks. The humps and spikes are often the points of

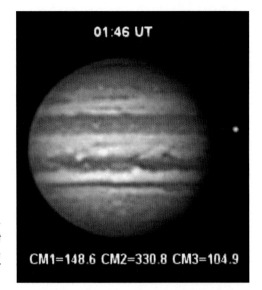

01:46 UT

CM1=148.6 CM2=330.8 CM3=104.9

Fig. 3.14. In 1998, the bluish-gray projections and festoons were very prominent on the NEBs. Also note the fragmented, discontinuous STB. Jupiter on November 23, 1998. (Credit: Donald C. Parker).

01:59 UT

CM1=87.8 CM2=76.6 CM3=24.5

Fig. 3.15. Prominent bluish-gray 'plateaus' on Jupiter's NEBs. Note the narrow NEB with a bright rift. The GRS has a very small, intense orange center. Jupiter on May 22, 2004. South is up. (Credit: Donald C. Parker).

departure of gray wisps or festoons, some of them most delicate and some quite conspicuous, that seem to issue from the S. edge of the belt and look as if they were dispersing like smoke in the Equatorial Zone. Frequently, however, they do not simply vanish but curve round (no apparent motion is implied by these attempts at simile) and return to the belt, almost certainly reaching it at a point where another projection appears. Sometimes a wisp will curve right over one projection and return to the second one following its point of departure. Quite often one of them will fork into both preceding and following directions; this may lead to the formation of a series of gray arches with light, or even bright, central regions, the whole presenting a most fascinating spectacle, and any such light oval area may contain a bright nucleus, which, however, is seldom to be found near its center but more probably fairly close to the belt near one of its ends. If a projection has been by-passed by one of the curving wisps, the enclosed light area will not, of course, be elliptical but will assume a shape resembling that of a kidney bean [16]."

While these bluish-gray features, or festoons, are almost always present, the shape and characteristic of each individual feature can be subject to change over short periods of time. This morphology is of great interest. Peek observed in 1941 a projection that grew enormously in just 2 days [17]. Shapes and sizes of the features are so prone to change that it can be difficult to keep track of them reliably. Certainly, from one week to the next, these projections can change in shape to the point that an observer who does not carefully track them will have trouble reliably identifying the same feature the next week. From 1959 to 1964, the Association of Lunar and Planetary Observers (A.L.P.O.) made a special effort to follow these features close to solar conjunction. Thus, during that time it was possible to track and recover individual features from one apparition into the next, proving that they could exist for long periods of time. However, in recent history these projections and festoons have not been so reliably followed.

The probe from the Galileo spacecraft actually descended through one of these bluish-gray features and detected a lower than expected level of water vapor [18]. But, to the visual observer, they appear as dark features, trailing back into the Equatorial Zone (EZ). Often, bright ovals are seen immediately following these festoons, seemingly imbedded in the hook of the festoons themselves. The vertical structure of these bluish-gray features is quite fascinating and will be further discussed in Chap. 4.

To the untrained observer, the equatorial zone (EZ) can appear quite uneventful. Yet, there is actually much to watch for here. During 1999, the EZ was very active with projections and festoons from the NEBs intruding prominently into the zone. On November 25, 1999 I noticed that the EZ was actually darker than the north polar region (NPR) and almost as dark as the south polar region (SPR). On August 05, 2000 it was noted by many observers, and confirmed on CCD images, that the EZ was greatly disturbed, giving the EZ an overall dusky appearance, such that the south tropical zone (STrZ) was actually the brightest zone on the planet (Fig. 3.16). These disturbances can be seen from time to time.

Visually, the EZ most often appears as the brightest feature on the planet, cream or alabaster in color. This was the case during the apparitions of 2000–2001, 2001–2002, 2002–2003, and 2003–2004. However, there are exceptions. Closer examination usually reveals there is more going on here. During Jupiter's history, the EZ has occasionally taken on a darker, yellow-ochre to brownish color, or sometimes a dusky appearance as the EZ undergoes a coloration event. Sometimes this effect can be subtle.

During the 1999–2000 apparition the EZ was bright and the festoons coming off the NEBs were very intense and prominent. Several of these festoons could be seen at various locations around the planet. The space between these festoons was generally bright and alabaster in color. However, by September 2001 the festoons were no longer so intense and the color of the EZ had begun to change. During 2000–2001 the EZ underwent a coloration event that would last beyond 2004.

By September 2001, the EZ coloration event gave a curious appearance to the planet. Compared to 1999, the decline of the projections and festoons gave a somewhat "clean" appearance to the EZ. These normally prominent features were conspicuous by their lack of intensity! The EZ had taken on a yellowish to yellow-ochre coloration for fully three-quarters of the distance from the NEBs to the SEBn.

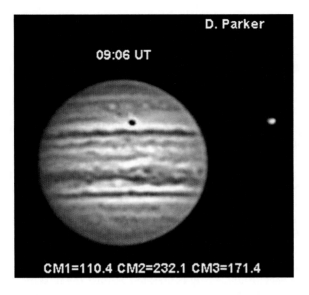

Fig. 3.16. Jupiter with a very disturbed EZ. The NTB is prominent with a couple of small projections on the southern edge. The northern edge of the NEB is withdrawn and undulating, and the SEB is split into two components. One of the Galilean moons is casting a shadow on Jupiter's globe. August 5, 2000. South is up. (Credit: Donald C. Parker).

Fig. 3.17. A low intensity GRS followed by a faded SEB. Note the coloration event in the northern two-thirds of the EZ. Jupiter on January 15, 2003. South is up. (Credit: P. Clay Sherrod).

During this time, due to much turbulence following the GRS, the SEB following the GRS for several degrees had faded (Fig. 3.17). The GRS itself had also declined in intensity. This fading of the SEB and GRS, and the dusky appearance of the EZ, combined with the coloration of the EZ, made it difficult to make out visually where the GRS/RSH ended and the SEB began! For several degrees following the GRS, the EZ was actually more intense than the SEB! It was still quite easy to discern the GRS against the SEB in CCD images. However, as late as December 2002 the GRS and

SEB following it were still faded. During the 2002–2003 apparition the northern three-quarters of the equatorial zone continued to display a light yellow-ochre coloration. A CCD image by Eric Ng taken on May 01, 2003 revealed an EZ that was almost entirely colored with a pronounced yellow-ochre appearance (Fig. 3.18). Only the very southern edge of the EZ was bright, like a thin bright line extending all around the planet just north of the SEB. By 2004, the NEBs projections and festoons were again easily seen. These features, along with the continued coloration of the EZ, and the absence of the NTB and NNTB gave the appearance of the NTrZ and NTZ being the brightest regions of the planet.

Sometimes the EZ will display an equatorial band (EB). The EB is most often seen as a thin, bluish-gray band or line running through the length of the EZ. It is not always present, and when present it may be broken or discontinuous. Often, when the EB is present at the same time as the bluish-gray projections of the NEBs, the projections will be seen to loop into the EZ and tie into the EB (Fig. 3.19).

During 2006, the projections and festoons trailing into the EZ were so prominent as to give the EZ a dusky appearance visually. The festoons trailed back into a bluish-gray equatorial band (EB). The area between the EB and the northern edge of the SEB was very bright, easily contrasted against the duskiness of the rest of the EZ. Many of the bluish-gray projections/festoons were accompanied by a bright oval, or porthole, following the projection and tucked u next to it. Some of these were bright enough to see visually. The festoons, portholes, and the various shadings of gray gave the EZ a very stormy appearance, especially visually through an eyepiece. Because of this, the NTrZ and the STrZ both appeared brighter than the EZ (Fig. 3.19).

In the SEBn, especially near the longitude of the GRS, extensive turbulence can be witnessed for several degrees of longitude preceding and following the GRS, as just mentioned. This turbulence is especially prominent following the GRS, where

Fig. 3.18. Jupiter on May 1, 2003. Notice the yellow-ochre coloration of the northern two-thirds of the EZ. The NTrZ was the brightest zone on Jupiter at this time. Webcam image with south up. (Credit: Eric Ng).

D. Parker
Coral Gables, FL
Seeing good: 7-8
Trans fair: 4
Wind E 6-10 kts.
Alt: 39 degs.

18 Aug 2006
00:21:55 UT

16-in Newt @ f-22
SKYnyx 2-0 camera
Astrodon Filters:
R = I Series;
G,B = E Series
RRGB. 34 fps

CM1=246.3 CM2=114.2 CM3=281.2

Fig. 3.19. Jupiter with prominent festoons trailing into the EZ. Notice at this late date the NTB and NNTB are still absent. August 18, 2006. South is up. (Credit: Donald C. Parker).

Fig. 3.20. Jupiter on March 8, 2003. Note the wake following the GRS causing a disturbed SEB. South is up. (Credit: Ed Grafton).

white ovals can appear in the south equatorial belt (SEB) (Fig. 3.20). These bright ovals can often be seen trailing back from the GRS toward the following side of the planet for several degrees of longitude. This bright area of continuous turbulence can vary in intensity, but is usually easy to make out against the usually darker, reddish-brown color of the SEB itself.

Sometimes, small bright ovals can be seen in the SEB near its northern edge preceding the GRS. There have been several instances of small, bright ovals in the

SEBn that have approached and passed north of the GRS, sometimes being caught up in the currents surrounding the GRS and swirling around the edge of the GRS into the middle of the SEB. Sometimes these bright spots will actually appear to enter into the GRS itself, offering an opportunity to track the anti-cyclonic behavior inside the GRS. This phenomenon was well observed on CCD images of Jupiter taken during September and October 2002.

3.1.6 The South Tropical Region

The South Tropical Region extends from latitude 9° south to 27° south. The South Tropical Region is considered by some to be the most interesting region of the planet. This is no doubt due to the fact that the GRS resides in this region, situated on and actually protruding into the southern edge of the SEB.

The South Tropical Zone lies just south of the SEB, between the SEB and the South Temperate Belt (STB). This zone, like other zones on the planet, normally displays a bright cream or alabaster appearance. Sometimes shadings can be seen in the zone, and disturbances and dislocations can also be seen from time to time. Markings in this zone are often difficult to see visually although easily detected by CCD or web cam imaging.

A further discussion of the SEB is appropriate here, as this is where many of the cyclical events on Jupiter occur. In addition to being the home of the GRS, the South Tropical Region is home to the famous SEB fadings, SEB Revivals, partial fadings, mid-SEB outbreaks, South Tropical Disturbances (STrDists), and South Tropical Dislocations.

The SEB is often wide and dark like the North Equatorial Belt. Sometimes the SEB is slightly wider. Normally, the SEB displays a reddish-brown color, just as the NEB does. At times, the SEB can split into two parts, a northern and southern component. When this occurs, the bright zone between these two components is referred to as the SEB zone (SEBZ). In November 1998 the SEB was a solid belt at longitude 292° System 2, with a very narrow belt of turbulent white clouds running along the northern edge of the belt. However, at 236° System 2 in September 1998 the northern half of the SEB was very bright, with turbulent white clouds. Later, during September 2000, Jupiter's SEB was split into two components, with the northern component, SEBZ, and southern component occupying the width of a normal, solid SEB (Fig. 3.21). The southern edge of the SEB (SEBs) was uneven, with bright bays protruding into the southern edge. Through February 2002 and beyond, the SEB continued to be divided into this double component. Although faint, in May 2002 the northern and southern components displayed a slight reddish-brown color, with the northern component the darker of the two. By February 2003, the SEB was again solid along most of its length, except for the usual bright, turbulent region following the GRS. During 2006, the SEB was again broad with a general reddish-brown color. As usual, there was much bright turbulence following the GRS. In April 2006, there was a really large white oval in the wake of turbulence following the GRS.

Although the SEB does not form large condensations or barges like the NEB does, sometimes small, reddish condensations or spots can form on or near the southern edge of the SEB. These very tiny versions of the GRS, while sometimes difficult to make out visually in a small telescope, can provide us with an opportunity to track the drift rate and currents at this latitude of the planet.

Fig. 3.21. Jupiter on September 6, 2000 with SEB split into a northern and southern component. Note STB fragment. South is up. (Credit: Donald C. Parker).

CM1=334.4 CM2=219.4 CM3=167.0

I think one of the most dramatic and interesting phenomena to observe on Jupiter is an SEB Revival. An SEB Revival is actually a series of events, the cause of which is not fully understood. During an SEB Revival, the SEB will fade, taking several months to completely disappear. The fading will normally begin in the southern component. The northern component does not always disappear; but it will often fade completely away. Once the SEB has faded, the GRS will often darken and become more intense. Then, after remaining faint for one to three years, the Revival will begin. The Revival starts at a single point in the belt, and from this point dark and bright spots begin to appear. From this point source spots are carried along with the currents of the SEB. Other spots well up from the point source and are carried along in the SEB currents, spreading out across the width of the belt. Eventually the material from these spots reaches the GRS. When that happens the GRS fades. The SEB continues to be filled with the dark and bright material and is finally restored. At present, a complete SEB fading followed by a Revival is well overdue. When it happens, chances are an amateur astronomer will be the first to notice the fading and report it.

Another SEB phenomenon is the Mid-SEB Outbreak. A Mid-SEB Outbreak in appearance mimics the beginnings of an SEB fading. When a Mid-SEB Outbreak occurs, there is an apparent fading of a portion of the SEB for several degrees of longitude. However, the remainder of the belt remains visible. The typical Mid-SEB Outbreak first appears as a small white streak or spot in the southern portion of the SEB, removed from the longitude of the GRS. This spot stretches out toward the following direction of the belt with other spots continuing to develop at the place of origin of the first spot. Soon, the SEB is covered with a very turbulent region of bright and dark streaks and spots resembling the continuous turbulent area following the GRS. This condition of the SEB can exist for several months. A Mid-SEB Outbreak occurred in 1998 and was easily observed, both visually and with CCD imaging. In 2006 there was another SEB outbreak that encircled much of the planet. As it progressed around the planet, it gave the SEB the appearance of being split into two components, a northern and southern component (Fig. 3.22).

The STrDist is another interesting phenomenon. Rogers describes a STrDist as "a coherent sector of shading and disturbance that spans the South Tropical Zone (STrZ) [19]". This zone resides between the SEB and the STB. The disturbance is characterized by a dark coloration or shading in the STrZ. This dark coloration is initially short in length, concave on the following and preceding ends, and usually bounded on each end by a small white oval. Often, the disturbance begins near the preceding edge of the GRS. The Disturbance normally fades when the SEB fades prior to a SEB Revival. Disturbances normally last several months [20].

South Tropical Dislocations are also seen in the STrZ. A South Tropical Dislocation occurs when a faded section of the STB passes the GRS. When this happens the STrZ preceding the GRS darkens, so that the normal pattern of belts and zones is reversed [21]. According to Rogers, in the 1970s a sector of the STB turned white, and since then this 'STB Fade' has repeatedly induced structures like STrDists preceding the GRS, giving rise to a dislocation of the normal pattern of alternating belts and zones. Moving with the South Temperate Current this STB Fade passes the GRS every two years. On some of these passages, it has induced a darkening of the STrZ on the preceding side of the GRS, usually in the form of a South Tropical Band although additional shading may be present. The term 'South Tropical Dislocation' has been applied to this combination of the white STB and dark STrZ, and also to the remarkable suite of structures that develop from it [22]. Although the STB itself in recent years has often been difficult to make out visually, including any STrDist, the dislocation is fascinating and valuable to observe. CCD and webcam images normally show these events quite well.

Fig. 3.22. Jupiter with a prominent Mid-SEB Outbreak underway on March 2, 2006. South is up. (Credit: Donald C. Parker).

3.1.7 The Great Red Spot

The GRS is no doubt the most recognizable feature on the planet. It is embedded in the southern edge of the SEB and is nestled between wind jets to the north and south [23]. It is an anti-cyclonic storm, rotating in a counter clock-wise direction, not too dissimilar from hurricanes on Earth, only incomprehensibly larger (Fig. 3.23). According to Schmude, the longitudinal length of the GRS was 23,000 km during the 1992–1993 apparition. Since Earth is 12,756 km in diameter, almost two of them could fit inside the GRS! The GRS varies in color and shape with time. Sometimes, the GRS is a vibrant red against a cream-colored background [24]. At other times, such as is apparent during South Equatorial Disturbances, the spot seems to fade and blend into its surroundings [25]. Although the GRS can be quite dark, since 1997 it has been rather faint, with a color of salmon-pink or light orange reported by most observers instead of deep red. As a consequence, the GRS has been difficult to observe in recent years, especially by amateurs new to Jupiter observing. Of late the GRS has been in the shape of an oval, or ellipse. During some past apparitions, the preceding and following ends of the GRS were more pointed.

A dark, oval spot near the latitude of the present day GRS was first observed by J-D Cassini and others beginning in 1665 [26]. Observed for a long, long time, there is some conjecture as to whether or not this weather system has persisted for over 350 years [27]. Indeed the system currently identified as the GRS can only be tentatively traced back to circa 1830, 120 years after the last sighting of Cassini's spot [28]. In 1878, the present day GRS was observed to be a deep, brick-red color [29]. Since that time, the GRS has often been very faint or totally faded. The mostly cloud-free area around the GRS is traditionally called the GRS "hollow" [30], or simply Red Spot Hollow. The Red Spot Hollow (RSH) marks the position of the GRS when the GRS is faded. The Hollow protrudes distinctly into the southern edge of the SEB.

From apparition to apparition, the color of the GRS can change noticeably. In fact, past changes in the STrZ and SEB have demonstrated that the GRS and its surroundings can change on very short time scales of less than five months [31]. Historical notes about the physical appearance of the GRS and its surroundings also indicate variability of both very short and long time-scales [32]. In general, the GRS is never "geranium" red, nor completely white [33], but no obvious periodic color trend is evident. It is observed, however, that when the GRS "ingests" smaller white vortices, the northern portion or the so-called "eyelid" of the GRS does whiten [34]. Occasional brightening in the northern part of the spot, is often associated with the ingestion of smaller eddies that are sheared apart inside the storm or with interaction with the South Tropical Zone [35].

Fig. 3.23. Jupiter's Great Red Spot. Note the darker condensation in the center of the GRS and the overall salmon-orange color, surrounded by a slightly darker collar of material. The GRS is nestled into the Red Spot Hollow. A bright gap can be seen north of the GRS, between the GRS and the southern edge of the SEB. Bright turbulence can be seen in the wake following the GRS. South is up. (Credit: Ed Grafton).

During the 2000–2001 apparition, the GRS was faint, with a darker condensation in its very center (Fig. 3.23). This dark center was difficult to observe visually, but it was easily made out in CCD images. The GRS was bordered on its southern edge by a thin gray line. Sometimes, small white spots can be seen entering the GRS. When this happens we have an opportunity to monitor the circulation inside the GRS itself. According to Schmude, during the 2001–2002 apparition, the GRS was reported to have no color or a faint orange-pink color [36]. By the beginning of the 2003–2004 apparition, many observers were reporting a slight darkening of the GRS with a slightly more intense pink color. This appearance continued through 2006, with the GRS displaying a moderately intense orange-salmon color, and a darker condensation in its center.

In the past, small dark condensations have also been observed to circulate around the GRS, leading to the conclusion that the GRS is a vortex. These observations helped to identify the circulating currents inside the GRS. During the apparition of 1965–1966 a dark spot was observed on the north edge of the STB that approached the GRS in decreasing longitude. Upon reaching the GRS it continued to pass the GRS on the south side of the GRS. After reaching the preceding side of the GRS it did not pass it, but continued around the GRS to the northern side of the GRS and returned to the following end of the GRS. The spot was photographed by Elmer Reese and B. A. Smith of the New Mexico State University Observatory, and observed by them from December 1965 to January 1966. The spot circulated completely around the GRS in 9 days [37]. This observation was a significant contribution by amateurs. The Voyager spacecraft also acquired images of spots that allowed scientists to further study these currents. Dark spots were observed as recently as the 2001–2002 apparition [38].

To say we do not know everything we would like to know about the GRS is an understatement! We do not possess a complete understanding of the cause of many of the phenomena seen on the planet nor do we know with certainty when events will occur or reoccur. However, we do know that certain events seem to follow others. For example, the GRS alternates between periods when it is dark and faint. When SEB Fades occur, the GRS often darkens, only to fade again during an SEB Revival as the SEB is restored to its normal appearance. As this is written, a complete SEB Fading is long overdue on Jupiter. The observation of such an event is of great importance and some diligent amateur will probably be the first to note the start of such an event. When it happens, will the GRS darken as it has done in the past? And how much time will pass before the SEB Revival begins? Any inconsistency from previous patterns might indicate a change in Jupiter's weather patterns. Regardless, any new behavior would be of extreme importance to the professional community.

The size of the GRS has been changing over the years. According to Simon-Miller et al. the latitudinal extent, or north–south distance, of the GRS has remained constant, with ground based observations from 1880 to 1970 indicating a latitudinal length of 11° ± 1°. This was consistent with Voyager data indicating a measured latitudinal extent of ~12,000 km [39]. However, the length of the GRS, longitudinally, has been diminishing over the years. In 1882, the GRS had a length of 34°. Until 1920, its length remained 30°, but since then it has never been as long again [40]. It has been known since the early 1900s [41] that the GRS is shrinking in apparent longitudinal extent.

If the latitudinal extent of the storm is thus assumed to be constant, the aspect ratio can be plotted for the older observations, indicating that the GRS is indeed becoming more round in appearance [42]. The system currently has an aspect ratio of about 1.6 and, obeying the historical trend, the GRS should become round by the year 2080, although that is not a stable configuration [43]. Simon-Miller, et al. found a longitudinal shrinkage rate for the GRS of −0.193° per year. At this present shrinkage rate, the GRS would be approximately round by the year 2040 [44].

The accuracy of any measurement of the length of the GRS is dependent upon the observer's ability to accurately identify the preceding and following edge of the system. This may seem straightforward but is not necessarily so. Poor seeing can undermine ones ability to see fine detail and transit timings under poor conditions can induce error into the timing. Likewise, measurements of longitude taken from CCD or webcam images can also suffer. Crisp, focused images of high resolution are needed to get accurate and reliable measurements. Sometimes the preceding and following edge of the GRS can be mistaken for the preceding and following edge of the Red Spot Hollow, even when measuring CCD or webcam images. Prior to the advent of CCD cameras and webcams, measurements of the GRS could only be made with the use of filar micrometers or central meridian transit timings, to be discussed later in this book. In spite of the inherent inaccuracies, multiple measurements by a large number of observers are desirable, and can average out the errors of measurement.

To illustrate the shrinking of the GRS over the years, we can review the historical record. The difficulty of measurement is illustrated by the results reported by various observers and researchers. All longitudes are given in System 2. Well known Jupiter observer Elmer J. Reese determined a length of 24.4° ± 1° during the 1962–1963 apparition [45]. During the 1968–1969 apparition Phillip Budine reported a length of 23° [46]. Giancarlo Favero et al.. reported 22.9° ± 0.5° in October 1975 [47]. Phillip Budine reported a length of 20° during the apparition of 1967–1968 [48]. A length of 23° was reported by Phillip Budine in 1983 [49]. A length of 19° was reported by Phillip Budine for the 1985–1986 apparition [50]. For the 1986–1987 apparition Budine determined a length of 22° early in the apparition, and 25° late in the apparition [51]. Richard Schmude reported 23.6° ± 1° during February 1989 [52], while Lehman et al. determined a mean length of 20° for the 1989–1990 apparition [53]. Schmude determined a length of 25.5° ± 3° for the 1989–1990 apparition [54]. According to Lehman and McAnally, during the apparition of 1998–1999, the GRS presented itself as a much smaller feature inside the RSH. The GRS proper was determined to be only 10° in length, while the RSH was 25° in length [55]. Using only high quality CCD and webcam images in which the preceding and following edge of the GRS could be accurately identified, I measured the length of the GRS to be 19° in November 1999, 17° in September 2000, 20° in September 2001, 17.5° in January 2002, and 17° in late January 2003. During the 2003–2004 apparition I determined the mean length of the GRS to be 17° in longitude. During 2006, I measured the length of the GRS to be 16°. Observations and measurements of the GRS by astronomers, including amateurs, will continue to be very important.

Sometimes the GRS presents a curious appearance, not quite filling the entire space occupied by the red spot hollow (RSH). This appearance has been noted during several recent apparitions. One such apparition, the 1998–1999 one, has already been mentioned above. In a CCD image taken by Donald Parker on September 04, 2000 a salmon-orange colored GRS was nestled in the southern two-thirds of a bright RSH. The bay of the RSH protruded significantly into the SEB, with bright material filling a large gap between the southern edge of the GRS and the northern edge of the SEB. The GRS and RSH again presented this kind of appearance in a CCD image taken by Maurizio Di Sciullo on January 12, 2002 in an image taken by Eric Ng on January 28, 2003 in an image taken by Tan Wei

Leong on February 23, 2003 an image by Cristian Fattinnanzi on April 02, 2004 and on webcam images taken by Parker on May 06, 2004 and 22. Sometimes the color in the GRS can appear to be concentrated into a small area inside the GRS itself, giving the GRS an even smaller appearance. This was the case on a CCD image taken by G. Kiss on November 18, 2001 when the color of the GRS was concentrated into an ellipse in the southern half of the GRS. This effect was again seen in a CCD image taken by Parker on January 21, 2002. In all the images cited above, an even darker concentration of color was seen in the center of the GRS, giving the appearance of a dark spot inside the GRS! This has been a common appearance in recent years. By April 2004, images by Fattinnanzi, Parker, and others revealed a GRS in which the color again filled the entire ellipse of the spot; however, the dark center remained.

The source of color for the GRS is not understood. However, it appears that Jupiter's visible color is contained in the tropospheric haze [56], which is an extended, colored, tropospheric haze denser than that seen on the rest of the planet, and reaching higher in altitude than most other locations on the planet [57]. Spacecraft missions have confirmed that the GRS has the vertical characteristic of a tilted pancake [58].

Much about the GRS is left unanswered, such as what created it, what maintains it, what colors it, and why there is no counterpart to it in Jupiter's northern hemisphere. And so, we do not know everything we would like to know about the GRS by a lot! One of my favorite endeavors is to keep a record of the changing appearance of the GRS. Certainly the GRS will continue to attract the attention of amateurs and professionals alike.

3.1.8　The South Temperate Region

The South Temperate Region generally extends from latitude 27° south to 37° south. In the past, this region has displayed some of the most interesting features on the planet.

The STB can be thought of as the southern counter part to the NTB. Like the NTB, it is a relatively narrow belt and usually not nearly as dark as the SEB. Its coloration is often described as gray, but I often note a touch of reddish-brown in the belt also. During the apparitions of 2001, 2002, and 2003, the STB was faint and some segments of the belt were totally faded, or at best, broken and fragmented (Fig. 3.21). Visually, features in this belt can be difficult to observe, as often the contrast between the belt and the surrounding zones can be very subtle. But, on occasion, a feature of some prominence can be seen. This belt can be continuous, broken into segments, or faded and completely absent. Dark spots have been observed here as well as bright ovals. According to Rogers, the STB was always quite distinct, at least until the 1980s [59]. However, since 1997 I have found the STB to be very subtle and difficult to observe visually. When seen, the STB has often been a narrow, faint gray belt, broken and discontinuous.

In 1998, the STB immediately following the longitude of the GRS presented a mottled but otherwise continuous appearance. By 1999, many segments of the STB were quite faded. In 2002, the STB was so faded as to be mostly absent, punctuated by short, darker segments in a few locations around the planet. By the end of 2002, the belt was so faded as to make it difficult visually to distinguish the STB from the northern edge of the SPR, since the South–South Temperate Belt (SSTB) was also very faint or absent. This condition continued into 2003. However, in 2003 a dark segment, approximately 50° long, was seen for many months. This dark segment was so prominent as to be easily seen visually in telescopes as small as 8-inches aperture. CCD images revealed that this segment was not solid, but composed of a string of condensations close together. In February 2003, this dark segment was positioned immediately south of the GRS, giving the impression of an eyebrow over the GRS (Fig. 3.24). The appearance of this segment allowed transit timings and longitude measurements to be made, providing a welcome opportunity to track the drift rate at this latitude. By July 2004, the STB was again presenting a mostly continuous, thin, although faint belt around most of the planet. I found this to be so even visually in an 8-inch telescope. Some segments of the belt continued to be faded. In 2006 the STB was absent or at least very difficult to see around most of the planet, being most easily seen following south temperate oval (STO) BA. In some locations where the STB was faded, a very light remnant could sometimes be made out.

As mentioned, short dark segments and condensations or "spots" may occasionally be present in the STB. As recent as 2003, a very short, dark segment or condensation was observed on the southern edge of the STB (STBs). This segment was followed for several months, and was seen visually as well as being detected on CCD and webcam images. A darker segment was seen for a short distance following oval BA during 2006. Segments and spots like this are not seen during every apparition. However, one of the most prominent spots ever seen in the STB appeared in 1998.

The South Temperate Dark Spot of 1998 was a fascinating feature to observe. While there is evidence of this spot on CCD images by Miyazaki in 1997, the spot did not come to notoriety until 1998 (Fig. 3.25). As the Assistant-Coordinator for Transit Timings of the A.L.P.O. Jupiter Section, I first received a report of the spot on 1998 June 20 from Harry Pulley, who had observed the spot and recorded a transit timing on that date. The spot was subsequently observed and reported by other observers over the next several weeks and I initially determined the spot had a drift rate of −5° per 30 days. On July 22, 1998 I issued an alert message from A.L.P.O. via the Internet and gave this feature the provisional name of STB Dark Spot #1. Visually, the spot appeared as a very small dark spot, like a pencil spot on

Fig. 3.24. Dark segment of the STB, giving the appearance of an 'eyebrow' south of the GRS on February 18, 2003. The STB is discontinuous. The NEB is narrow and the NTB is absent. Note the five bright ovals in the SSTB. Note also the coloration of the northern two-thirds of the EZ. South is up. (Credit: Ed Grafton).

Fig. 3.25. Jupiter on August 3, 1998 and the South Temperate Dark Spot of 1998. Also seen is STB Dark Spot #2, an elongated spot following close behind. Two major ground telescopes and the Galileo spacecraft made investigations of the 'dark spot'. South is up. (Credit: Donald C. Parker).

a page, almost as dark as a moon shadow. Being so small, the spot was difficult to see visually, but once found it seemed to jump out at the observer. The visual effect was remarkable.

After July 14, 1998 the drift rate of the spot increased to −15° per 30 days, prompting me to issue another alert message. This second message caught the attention of Dr. Glenn Orton, of the Jet Propulsion Laboratory, who requested additional information. Dr. Orton was one of the Galileo Spacecraft Mission Interdisciplinary Scientists. Dr. Orton and I began corresponding on a regular basis. From the date the first alert message was issued until the end of the 1998–1999 apparition, the spot was observed continuously around the world. In fact, world-wide participation was so intense that there were few occasions that a transit of the spot was not observed at least twice a week! There was tremendous cooperation, especially among amateurs; and the interest by the professional community was a great example of professional-amateur cooperation!

On September 21, 1998 I issued another alert over the Internet announcing a second dark feature had been seen in the STB. This feature was given the pro-visional name of STB Dark Spot #2. Dr. Orton informed me that a decision had been made in the professional community to examine Dark Spot #1 more closely. It was Dr. Orton's intention to observe Jupiter at 5-μm with the NASA Infrared Telescope Facility (IRTF) Telescope and Near Infrared Camera (NSFCAM) from Mauna Kea for three full nights starting on or about September 27, 1998. At the same time, Dr. Terry Martin was to observe from Palomar Observatory using a middle-infrared camera and a spectrometer. Dr. Orton anticipated they would also capture the feature with 27-μm observations made with the Galileo Spacecraft Photo-Polarimeter-Radiometer (PPR) sequence on Galileo's orbit number 17 [60]. Dr. Orton later informed me that all three observations were successful. You can imagine how excited we were in the amateur community, having contributed to the professional interest in this feature that led to the allocation of observing time with such important resources!

These observations revealed that the feature was not a spot at all, but a hole in Jupiter's cloud tops. The "spot" was bright at 5-μm (the infrared). Since infrared detectors detect warmth, this brightness indicated the spot was a warm feature. In other words, we were detecting the warmer temperature below Jupiter's cloud tops escaping up through this hole. In visible light, this hole was dark, since we were seeing through it to lower cloud decks.

This "spot" was observed for two apparitions with great success, and caught the attention of amateurs and professionals alike. Not only was this an exciting feature to observe and record, the international cooperation that ensued was a great accomplishment for astronomers [61]. When features like these are seen, Jupiter observers are presented with an opportunity to monitor the drift rates in this belt with great accuracy.

The features most prominently seen on the STB are white ovals [62]. Many of these ovals are quite small and difficult to observe visually due to the low contrast of the ovals against the zones on either side of the belt. The most famous STO today is the larger oval BA (Fig. 3.26).

Oval BA is the loan survivor of three ovals that originated in the South Temperate Zone (STZ) between 1939 and 1940 [63]. The history of these ovals is quite fascinating, and the eventual demise of two of them presented astronomers, professional and amateur alike, with an adventure in observational astronomy we shall not soon forget.

The three STOs began their existence as bright sections of the STZ. During 1939–1940, three bright segments or spots appeared in Jupiter's south temperate region. The spots rapidly contracted during the 1940s to form three large white ovals by 1950, eventually shrinking to sizes of ~ 10,000 km by the 1990s. During their lifetimes, the ovals demonstrated wandering motions (movement in longitude), with 'mutual close approaches without mergers' [64].

Fig. 3.26. The bright South Temperate Oval 'BA' in the STB near Jupiter's central meridian on April 22, 2005. Note how the color of oval 'BA' is very close to that of the neighboring zones on this date. A small bright oval is in the SSTB just south of 'BA'. South is up. (Credit: Damian Peach).

According to Rogers, the three 'proto ovals' appeared when the STZ became sub-divided by the gradual appearance of three dark features, which initially appeared either as dusky sections of the zone or as segments of a south component of the STB. These dark features expanded longitudinally until they confined the three intervening bright sectors of the STZ into gradually contracting ovals. Elmer Reese had originally named the dark segments AB, CD, and EF. Subsequently, the bright ovals formed between them became known as BC, DE, and FA [65]. The STOs are anticyclonic like the GRS, rotating counterclockwise.

Voyager images in 1979 revealed the STOs to be anticyclonic vortices similar dynamically to the GRS. In 1998, two of the ovals BC and DE merged into a larger one, later designated BE, while Jupiter was too close to the Sun to be observed [66]. At the beginning of the 1998–1999 apparition, it was discovered that one of the three ovals was missing! Apparently, something unexpected happened while Jupiter was at conjunction, hidden from view behind the sun. The answer to this did not come immediately. Dr. Reta Beebe of NMSU and the International Jupiter Watch (IJW) contacted me seeking the longitudinal positions of the two surviving ovals. I had just assumed the duties of the Assistant Coordinator for Transit Timings of the A.L.P.O. Jupiter Section in 1997 and had the responsibility of keeping track of the positions of features in Jupiter's visible atmosphere. Dr. Beebe's intention was to use the Hubble Space Telescope (HST) to study the remaining ovals. In her words, no one was sure whether an oval had simply faded or two had merged. Eventually it was decided that the most likely scenario was that ovals BC and DE had indeed merged. The merged oval was renamed BE.

Previously the three ovals had drifted close to each other on many occasions, only to drift apart again. Often, smaller intervening cyclonic ovals were present, separating the larger anticyclonic ovals. It was always assumed that the dynamics of these systems would simply cause the three large ovals to bounce off of each other should they come together. News that two of the ovals had actually merged stunned Jupiter scientists.

As the 1999–2000 apparition began, the possibility of another STO merger, this time between ovals BE and FA, again presented itself. On April 30, 1999 the separation between the ovals' centers was only 18°. From April 30, 1999 to November 20, 1999 the distance between BE and FA varied in a dance that would see them drift closer together only to drift further apart again. The ovals drew closer then drifted apart several times. By November 20, 1999 the ovals were at or approaching conjunction with the GRS. Astronomers wondered if the passing of the GRS would change the ovals' drift rates. The ovals continued their dance of drifting closer together and then farther apart, but closing the overall distance over time. By January 2000, CCD images by the Pic du Midi Observatory revealed that a cyclonic cell between the two ovals had disappeared, which could make it easier for the two ovals to collide and merge (Sanchez-Lavega personal communication). On February 8, 2000 the ovals centers were measured as 12° apart. More significantly, the following edge of BE and the preceding edge of FA were only 5° apart! At this time as a staff member of the A.L.P.O., I issued an alert over the Internet, receiving immediate response from astronomers at JPL and Cornell University. By March 17, 2000 the ovals were in contact with each other but undisturbed. From this time forward, events developed rapidly. On March 19 and 20, 2000 oval BE shifted northward with FA overtaking it. Indeed, infrared images from the Infrared Telescope Facility (IRTF) on Mauna Kea, showed the bright methane cloud-caps over the ovals rotating around each other. On

March 21 and 23, 2000 CCD images showed BE was disrupted. By April 7, 2000 CCD images showed BE/FA as a single object, very diffuse [67].

And so, during a three week period beginning in March 2000, ovals FA and BE completed their merger in the south temperate region. The merger took place when the ovals were southeast of the GRS, and after the disappearance of a smaller clockwise rotating oval that had been between them. The high altitude oval clouds of ovals BE and FA demonstrated counter clockwise rotation around each other, then merged and began shrinking. The interaction of deeper clouds did not show mutual rotation [68].

Prior to the merger of the two remaining ovals BE and FA, HST images revealed that oval BE had a diameter of ~9,000 km and oval FA had a diameter ~7,700 km. Situated between them was a smaller oval, designated O1. Oval O1 had a diameter of ~5,000 km. In latitude, oval BE was located at −32.7°, oval FA at −33.6°, and smaller oval O1 was at −31°, or south of both STOs. Oval O1 was first observed in May 1998 and was similar to one observed between ovals BC and DE before they merged in 1998 [69].

Close interactions between these three vortices (BE, FA, and O1) began in November 1999 when the ovals passed the GRS. At that time oval O1 began a southward migration, moving from latitude −31° to −36°, "moving along an arc of 35° that encircled oval BE in the anticyclonic (counter clockwise) sense", according to Sanchez-LaVega et al. Oval O1 moved from the cyclonic region to the adjacent anticyclonic (counter clockwise) one [70].

It would appear that the removal of oval O1 allowed ovals BE and FA to interact directly. According to Sanchez-LaVega et al. "as the ovals approached each other, BE moved with a velocity of $u = 0.6\,ms^{-1}$, whereas FA decreased its velocity from u = 1.6 ms^{-1} in January 2000 to u = 0.9 ms^{-1} in early March 2000." On March 17, 2000 FA pushed BE northward by ~3.2° (4,000 km) as observed at high altitudes. Then on March 21, 2000 the pair began an anticyclonic (counter clockwise) orbit about each other. By April 3, 2000 the ovals had merged, and by April 14, 2000 had decreased in size to a compact state [71].

At middle and lower levels (altitudes) the interaction was a little different, and the first interaction occurred March 12–15, 2000. Between April 7 and 14, 2000 new oval BA was still forming with a double nucleus of 5° separation still seen. Later, on September 2, 2000 the area of BA was determined to be ~70% of the sum of the areas of BE and FA [72].

According to Sanchez-LaVega et al. the orbiting action observed in oval O1 and in ovals FA and BE can be explained by the presence of a velocity field induced by the vortices on each other. The observed interactions consisted of well-known phenomena that had been observed and predicted by models in computer simulations. It appears that most interactions of smaller scale spots observed in Jupiter lead to mergers. However, prior to 1998 the three STOs had always just bounced off of each other. In spite of their long history as separate objects, the two remaining STOs had collided and merged together. So, after 60 years the three ovals have finally coalesced into one vortex [73]. As of this writing, the surviving oval BA continues to thrive. With the last of the two STOs having merged, the remaining oval became known as oval BA [74, 75]. Like the GRS, oval BA is a huge counter-clockwise rotating vortice. Oval BA is still visible today, being especially well seen in CCD imaging. Visually, this oval can be difficult to make out due to its low contrast, especially if the dark collar of material that often surrounds it is absent.

Amateur and professional astronomers alike kept faithful watch over STO BA during the 2000–2001 apparition. In October 2000, oval BA presented a large, bright object. However, with a faded and broken STB, oval BA was often difficult to make out visually. A CCD image taken by Donald Parker on October 5, 2000 revealed a large oval BA surrounded by a collar of dark material, and trailed for a short distance by a dark segment of the STB. Without this dark material, BA would have been difficult to see even on CCD images, due to low contrast with the rest of the south temperate region. I used this dark segment of the STB to help pinpoint the following edge of oval BA when observing visually. However, by December 2001 the dark collar of material surrounding oval BA had almost vanished, leaving the contrast between the oval and its surroundings so subtle as to make it near impossible to visually make out the oval. Only low intensity dark material following the oval in the STB made it possible to see it at all. Even on CCD images by Ed Grafton in December 2001 and Maurizio Di Sciulla in January 2002, the low contrast of oval BA was startling. Eventually, some of the dark collar material returned. A CCD image by Eric Ng taken on February 22, 2003 revealed an oval surrounded by dark material, although oval BA itself was rather dusky and subdued, contributing to low contrast. By March 2004, oval BA was once again easier to see due to dark material surrounding it on its south, preceding, and following edge. CCD images by Rolando Chavez, Cristian Fattinnanzi, and Parker reveal this quite well.

Since they first appeared in 1939–1940, the STOs contracted and continued to contract during their lifespan. The three ovals were of different lengths longitudinally, with FA being the most diminutive of the three before the demise began. In the 1980s, ovals BC and DE were 8–9° in length while oval FA was 5° in length [76]. On March 11, 2004 I measured the length of surviving oval BA as 7°. The morphology, intensity, and drift rate of STO BA continues to be of great interest to Jupiter observers. Could the GRS have gone through the same process somewhere in its history? Keeping an eye on the behavior and condition of oval BA will be a valuable contribution that amateur astronomers can make, possibly for years to come.

Jupiter with the GRS and Oval BA
July 3, 2006 12:26UT
I: 270 II: 125 III: 280   S: 7/10 T: 3/5
© Christopher Go (Cebu, Philippines)

Fig. 3.27. Jupiter on July 3, 2006. South Temperate Oval 'BA' has turned red. Note how the color of 'BA' is very similar to that of the GRS. South is up. (Credit: Christopher Go).

With great surprise, oval BA changed color in late 2005 and became noticeably red in early 2006 (Fig. 3.27). Many amateurs began to refer to the oval as "Red Spot, Jr." I prefer to continue referring to this feature as oval BA, since this is what it truly is and because we need to preserve the continuity of our observational record. This change in the oval's color was first noted and reported by amateur astronomers, and by April 2006 professional astronomers had made a detailed study and analysis of this event. This historic change will be more fully discussed in Chap. 4.

It seems to me that, compared to the NTB, the STB is a much more active belt with a greater variety of features. What will the future hold? Only future observations will reveal how many ovals, spots, and belt fadings will appear. And what about oval BA? How long will it survive? Forever? That now seems unlikely. However, with no other large ovals to collide with, is there anything there to disrupt it? The south temperate region will continue to be one of the most important regions of the planet to patrol.

3.1.9 The South–South Temperate Region

The South–South Temperate Region generally extends from latitude 37° south to 53° south. Lying just south of the STB is the STZ, and just south of that is the SSTB. The south–south temperate region is exceedingly difficult to observe visually. To most observers this region appears as varying intensities of gray, with very few features that can be made out. Sometimes a dim, gray SSTB can be seen. Most often it is actually difficult to distinguish this belt, or the South South Temperate Zone (SSTZ) for that matter, from the northern edge of the SPR. Like the STB, the SSTB is a very narrow belt with a light gray coloration. While features here are difficult to observe visually, CCD imaging by amateurs has been quite successful. Normally, a number of small, bright ovals can be seen here, being of great interest to astronomers (Fig. 3.28). These ovals can provide our only chance to monitor the drift rates at this latitude. Unlike the much larger STO BA, these south STOs are quite small and of especially low contrast. Many visual observers have never seen one. Images with CCD cameras and webcams usually reveal them easily, as they stand out against the slightly darker gray coloration of the SSTB and SSTZ. Examination of CCD images during recent apparitions suggests that there are always six or more of these ovals each apparition. These ovals are of relatively small size, however, I did measure one oval as 5° in length on a CCD image taken by Cristian Fattinnanzi on April 02, 2004. Several ovals were seen, quite bright against the dusky, gray SSTZ. During 2006, several small bright ovals were again observed, being very prominent in CCD/webcam images (Fig. 3.28).

Like the NPR, the SPR often presents little that can be seen by amateurs, even with CCD cameras and webcams. Sometimes imaging can reveal very subtle, thin faint bands and barely noticeable zones marked only by very subtle shading differences. Sometimes small bright ovals are seen. Features that might offer a chance to track currents in this region are short-lived or very difficult to see. I did note during the 2003–2004 apparition what might be described as a very small south polar hood of bluish-gray color, which was also noticeable on CCD images (Fig. 3.29). However, the remarkable work by amateurs using CCDs and webcams caught many features that were of great interest. In fact, the SPR turned out to be much more active than the NPR! Small bright ovals that turned out to be long-lived were seen in the SPR at 60° south latitude, and another one was seen in the south–south–south temperate

Fig. 3.28. Four very bright small ovals are seen in the SSTB of Jupiter in this webcam image taken on April 2, 2004. The GRS can be seen emerging from the following limb of the planet. A prominent bluish-gray plateau with a trailing festoon can be seen on the south edge of the NEB. The northern edge of the NEB is very uneven. South is up. (Credit: Cristian Fattinnanzi).

Fig. 3.29. Jupiter with a very quiet SPR on February 10, 2004. Note the absence of any remarkable features in the region. By contrast, the EZ is very busy with prominent blue-gray features projecting into it from the southern edge of the NEB. Also note the subtle, yellow-ochre coloration of the northern two-thirds of the EZ. South is up. (Credit: Donald C. Parker).

current (SSSTC) at 50° south latitude. As technology advances and more tools become available to amateurs, the future may allow amateurs to do even more useful work in this region.

Hopefully, the past history we have studied here makes it clear that over a relatively short number of years, Jupiter can exhibit distinct and substantial changes in appearance. If you pay attention, you will never find Jupiter to be a boring planet!

3.2 Winds and Jet Streams in the Atmosphere

Features seen in Jupiter's atmosphere are not stationary, but drift with the currents and jet streams they are associated with. While the study of Jupiter's atmospheric dynamics is a whole discipline in itself, some understanding of Jupiter's atmospheric systems can aid the amateur in the study of Jupiter. When the drift rate of a certain feature is known along with its apparent latitudinal position, its association with a particular jet stream or current can be specified with some certainty. This knowledge will allow a prediction of its future position over time, allowing for the recovery and identification of the feature with some certainty following the end of an apparition or following a period of bad weather when there has been a lapse in observations. In more than one instance, the determination of the type of feature in a belt or zone was finally settled only after its drift rate had been determined and compared with historical data for the region being observed.

There are a number of currents and jet streams in Jupiter's atmosphere, some of which can only be observed by spacecraft. Here we will concentrate only on the ones that can be monitored visually. Following is a discussion of various jet streams and currents after Rogers, 1995 (Fig. 3.30).

The currents that can be monitored by visual observers fall into three categories. There is the great equatorial current, which encompasses the entire equatorial region. This current pro-grades, or moves in increasing longitude, at 7–8° per day, relative to System II. There are the nine slow currents, which govern most of the visible features outside the equatorial region, and have speeds of no more than 1° per day. And, there are the jet streams on the edges of certain belts, only observed

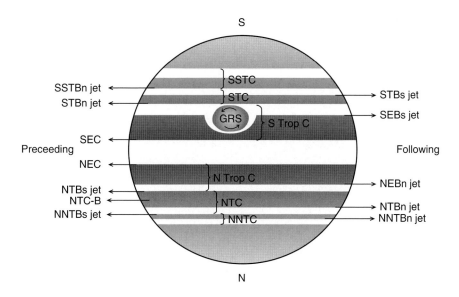

Fig. 3.30. A simplified depiction of the currents and jet streams on Jupiter, indicating the direction of flow in Jupiter's atmosphere. South is up. (Credit: John W. McAnally after Rogers 1995).

during infrequent outbreaks of small dark spots, which have speeds of several degrees per day [77].

The jet streams are detectable from Earth only when they are disturbed by distinct spot outbreaks. All the larger features move with the slow currents [78].

3.2.1 The Slow Currents

The slow currents had all been detected and recorded by 1901 (Table 3.1). Since that time, they have shown only temporary variations in speed. The cause for the difference in speed of the nine currents is not known. The speeds do not seem to be related to the types of features that are carried along by them, nor to any jet streams around them [79].

A variety of features can be seen in these slow currents, including bright and dark spots, medium and large spots, and cyclonic and anti-cyclonic spots. Regardless of the feature seen, they all seem to move in the same slow current. In some latitudes, either bright spots or dark spots are more common [80]. All latitudes of the planet present something of interest because of the variety of features and the currents

Table 3.1. Jupiter's jet streams and slow currents: latitudes and speeds

Jetstream	Latitude range B''	u (m s^{-1})	Change in longitude System II (° per 30 days)
SSSSTBn	53.4°S	+36.3	−129
SSSTBn	43.6°S	+42.1	−125
SSTBn	36.5°S	+31.6	−88
STBs	32.6°S	−20.8	+42
STBn	27–29°S	+44.3	−110
SEBs	20.0°S	−56.6	+117
SEC	7°S	+128	−276
NEC	5–8°N	+103	−224
NEBn	17.6°N	−24.3	+45
NTBs	23.8°N	+163	−375
NTBn	31.6°N	−32	+69
NNTBs	35.6°N	+34.5	−94
NNTBn	39.5°N	−14.8	+31
NNNTBs	43°N	+21.8	−68
NNNNTBs	48.2°N	+28.5	−94
NNNNNTBs	56.6°N	+14.1	−59
Current			
SSSTC	45–52°S	+0.1 ± 1.9	−8.3
SSTC	38–42°S	+6.6 ± 0.7	−25.1
STC	29–35°S	+7.5 to 1.5	−26 to −12
STropC	13–25°S	−3.6 (±1.0)	−0.1 (±2.3)
NTropC	14–22°N	+0.5 (±2.5)	−9.1 (±5.4)
NTC	25–34°N	−10.6 (±1.8)	+17.0 (±4.2)
NNTC	37–42°N	−2.9 (±1.2)	−0.3 (±3.2)
NNNTC	44–47°N	+2.5 (±1.4)	−15.2 (±4.1)
NNNNTC	49–55°N	−2.9 (±0.9)	+1.4 (±2.8)

B'' represents *zenographic latitude*; u (m s^{-1}) represents speed in meters per second in System III. (After Rogers 1995.)

with which they are involved. All features may not be present during each apparition, but there will be enough variety to maintain the serious observer's interest. Patience and attention to detail will enable the observer to gather significant data about the present condition of these currents.

The uniformity of the slow currents is important because bright and dark spots are now known to have opposite circulations, and actually lie on opposite sides of the retrograding jet stream, which are commonly deflected around them. The larger spots are therefore embedded in waves on the retrograding jet streams. The slow-current features observed from Earth are mostly large bright ovals or dark condensations of sizes comparable to the distances between jet streams [81].

3.2.2 The Jet Streams

Jupiter's jet streams are not too dissimilar to the jet streams on Earth. Both are fast moving currents. However, the jet streams on Jupiter appear to be more stable in latitudinal position that Earth's jet stream, which can vary in latitudinal position by a significant distance. Jupiter also has many more jet streams than Earth. These jet streams create some of the most exciting phenomenon in Jupiter's atmosphere. Certainly, over the years I have enjoyed watching the fast changing positions of features in these jet streams.

The fastest jet stream on the planet, which resides on the southern edge of the NTBs, was discovered during an outbreak of spots in 1880. Three other rapid currents on the SEBs, STBn, and NNTBs, were discovered during outbreaks in the 1920s. The fifth jet stream to be discovered from Earth, the SSTBn, was observed in 1988 [82].

Our view from Earth is selective, since the jet streams are only detectable during outbreaks. For the five jet streams that have been observed from Earth, latitudes and speeds determined by spacecraft agree with those determined from Earth. The latitudes of the jet streams, except for one, have not changed significantly over the period of Earth-based observations [83]. There are several other jet streams that have only been observed from spacecraft. However, for the five detectable from Earth, amateurs, especially with CCD and webcam imaging, can monitor and contribute significant drift rate data.

3.3 Color in Jupiter's Belts and Zones

Observation of color in Jupiter's belts and zones can be an aesthetically pleasing endeavor. Unfortunately, it also is one that can be very scientifically frustrating. Not only is contrast very subtle between features in Jupiter's cloud tops, but the determination of color by the human eye is difficult to quantify and fraught with subjectivity. Peek noted that the visual appreciation of color is such an individual affair that it can hardly be expected to furnish scientific data reliable enough to be used with confidence as a satisfactory basis for theoretical investigation [84]. Using the human eye, observers can only judge color in a relative and subjective manner, even if they are wise to possible instrumental and atmospheric illusions [85]. Nevertheless, I almost always make note of color in the record of observation when observing Jupiter.

A number of hurdles present themselves when attempting to observe the color of a celestial object from Earth. The type of telescope and optics is very important, and a reflector used in combination with high quality eyepieces is usually the better choice since it will minimize the effect of chromatic aberration.

Earth's atmosphere, or more specifically atmospheric dispersion, is a major source of trouble. Earth's atmosphere introduces chromatic effects that at low altitude are considerable and are particularly apt to vitiate color estimates in the case of a belted planet like Jupiter. The greater density of the air near the ground, compared with that at higher altitudes, not only brings about the well known effects of refraction but also causes the image of a point source, like a star, to appear as a short vertical spectrum with the violet at its upper and the red at its lower end. Under telescopic magnification this is especially noticeable, and in the case of a planet whose upper limb (lower in the telescopic image) may be seen edged in blue, while the lower limb has a reddish border. For a belted planet like Jupiter with bright zones, blue and red light will spill over from the edges of the bright zones and color the edges of the adjacent darker belts [86].

The use of color filters of known wavelength transmission can eliminate the subjectivity in determination of color. It can be noted that when observing Jupiter with a blue filter, the GRS appears darker, as the red light is not transmitted. Likewise, the GRS will appear bright when observed with a red filter. Filters can be used visually, or with photography and CCD and webcam imaging. Filters of known transmission are very useful scientifically and more will be discussed about their use later.

When describing color on Jupiter, it is probably wise to avoid thinking in absolute terms. The GRS is not red but reddish or reddish-brown, or salmon-orange, or pinkish. The belts are not brown or gray, but instead reddish-brown or grayish. There are shades of gray, shades of browns or ochre, colors tending toward something; however, there are no absolutes.

One of the most intriguing methods for removing subjectivity from the observation of color is put forth by Dr. Julius Benton. Benton is the Coordinator of the Saturn Section of the Association of Lunar and Planetary Observers. Visually made absolute color estimates should be carried out by comparing 'the planet' with a satisfactory color standard, according to Benton. Estimates of absolute color have to be made with instruments in integrated light (no filter). The A.L.P.O. Saturn Section has carefully investigated a suitable color standard, and it employs some 500 colored paper wedges for comparative use. One should have normal color vision for this type of work, and the color standard should (ideally) be illuminated by a tungsten lamp filtered with a Wratten 78 (W78) color filter [87].

In spite of the difficulties in determination of color, it continues to be important for observers of Jupiter to study and characterize color, including the patterns of change in coloration and intensity.

3.4 Summary

Some features may be seen to occur only in certain belts or zones. Features may be present during one apparition and absent the next. Some may be seldom seen, while others are almost always present. CCD images can reveal features difficult to make out with the unaided eye; or, even those that usually cannot be seen at all.

Some of Jupiter's features are so subtle that I know several experienced observers who have never made them out visually. It is important for the observer of Jupiter not to be disappointed if certain features seen previously are absent at a later date. Certainly, the absence of a feature, or type of feature, may turn out to be an important piece of data. Approach the planet with objectivity, without expectations, and every observation will be a new discovery. You will never be disappointed.

Chapter 4

Color, Chemical Composition of the Planet, and Vertical Structure of the Atmosphere

There is much about Jupiter that is not readily apparent to the amateur astronomer gazing through a small telescope in visible light. We can argue whether an understanding of such things should not simply be the domain of professional astronomers. Regardless, I believe well-rounded amateurs will want to learn as much as possible, and thus will appreciate the study of the planet all the more.

4.1 Color

Previously, we discussed the observation of Jupiter's colors. But, what causes the color in Jupiter's atmosphere? As we might surmise, that explanation is not so easily forthcoming. The different colors represent clouds with different compositions and different vertical structures [88]. The understanding of color in Jupiter's atmosphere is dependent upon some basic understanding of atmosphere. So, first, let us try to understand some basic principles about atmospheric structure.

Jupiter's atmosphere is divided into four main layers. These layers are the troposphere, stratosphere, thermosphere, and ionosphere (Fig. 4.1). The boundaries of the troposphere and stratosphere are determined by atmospheric pressure and temperature, both of which are interrelated, as temperature changes with height in atmosphere.

Jupiter emits almost twice as much heat as it receives from the Sun. This heat is emitted upward as infrared radiation. As this heat is radiated upward, it drives the changes in Jupiter's weather. Similar to Earth, temperatures in Jupiter's atmosphere change with altitude. Unlike Earth, Jupiter has no hard surface that we can see. On Earth we measure altitude from 'sea level.' Jupiter has no sea level, so here scientists make an assumption. On Earth we measure pressure in bars. Jupiter scientists use the same scale of measurement and assume that 1 bar is sea level for Jupiter. So that 1 bar of pressure is the same as an altitude of zero.

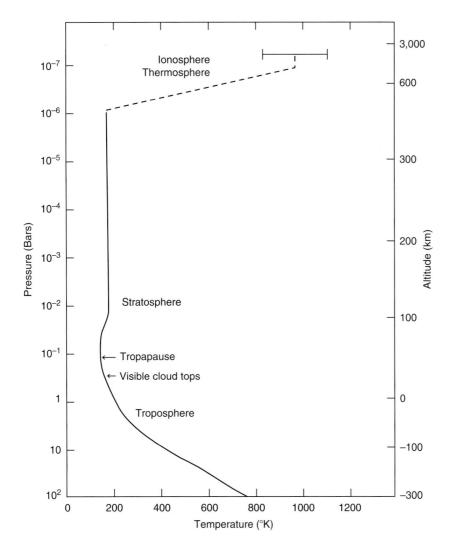

Fig. 4.1. A simplified model of the vertical structure of Jupiter's atmosphere, demonstrating how temperature and pressure vary with height. (Adapted from Rogers 1995.) [89].

On Earth, the lowest atmospheric layer is the troposphere, and so it is with Jupiter. On Jupiter, the troposphere extends from some low level to an altitude corresponding to the lowest temperature in Jupiter's atmosphere. This point is referred to as the tropopause; that is, the altitude at which the troposphere stops. Beyond the tropopause is the stratosphere. So, the tropopause is the boundary between the troposphere and the stratosphere. The tropopause is also described as the layer at which temperature stops decreasing. From the tropopause the temperature again rises into the stratosphere. So, at the tropopause, a temperature inversion occurs. Temperature increases with altitude above the tropopause. The next boundary layer is the stratopause. According to Amy Simon-Miller, "the stratopause occurs when the temperature stops increasing and begins to decrease again. So, the physical thickness and height of each layer is determined entirely by temperature." (Simon-Miller personal communication.)

Like Earth, the clouds that we see occur in the troposphere. In the lower troposphere, convection is the main method of moving the heat upward. In the upper troposphere, thermal radiation also plays a part. At the tropopause, convection almost stops [90]. Clouds generally cannot exist above the tropopause, except for the case in which strong convection causes the clouds to overshoot the tropopause a little and protrude into the stratosphere. The clouds in the troposphere are optically thick. We see the tops of the clouds illuminated by sunlight but we cannot see into the clouds. The cloud tops of the zones are generally higher than the cloud tops of the belts (Simon-Miller personal communication).

The stratosphere lies above the tropopause. In the stratosphere, heat is carried upward by thermal radiation, and there is no convection [91]. Temperature rises in the stratosphere with altitude. While the troposphere has distinct clouds, by contrast the stratosphere is composed of thin gases, sometimes interspersed or overcast with thin hazes or aerosols. Aerosols are small particles suspended in a gas. These gases are practically transparent, or optically thin. In the stratosphere, heat is absorbed from the Sun and from the radiation of Jupiter's own heat below. The stratosphere attempts to reradiate this heat upwards. However, the gases here are so thin, they cannot radiate heat away as rapidly as can be done at lower altitudes [92]. Thus, above the tropopause, temperature again rises with altitude, instead of falling as it does from 'sea level' up to the tropopause. At zero (0) altitude or 1 bar, the atmospheric temperature is ~165 K. The temperature at the tropopause is 105 K at a level of 100–160 mbar. The temperature at the upper stratosphere reaches about 170 K at 1 mbar. Here the temperature stabilizes until the thermosphere is reached at which point the temperature literally takes off! Beyond the stratosphere, the temperature in the thermosphere can reach between 850 and 1,300 K [93].

The ionosphere is defined as the upper layer of the atmosphere where there are many free ions and electrons. Hydrogen molecules are disassociated and ionized by solar ultraviolet radiation, and by electrons and ions raining down from the magnetosphere. The ionosphere extends several thousand kilometers above the molecular atmosphere. Here electrons can reach temperatures of 1,000–1,300 K. The thermosphere is defined as the upper layer of the atmosphere that is very hot. Here the neutral gas, 1,500–2,000 km above the cloud tops, can reach temperatures of 1,100 (±200) K. The thermosphere overlaps with the ionosphere [94].

We might suspect that color depends upon chemical composition; however, this is not necessarily the case. We refer to any agents that affect the shape of the reflection spectrum at continuum wavelengths in the extended visible spectrum from about 0.4 to 1 μm as chromophores [95]. To the question of color, spectra have not given any answers. Most potential colored substances (chromophores) do not produce distinct spectral lines, just very broad bands. Spectra of visible light show hardly any difference between the belts and zones [96]. All the major compounds predicted to form clouds in Jupiter's atmosphere are chemically simple and would be white. According to Rogers (1995), the main candidates for colored substances are alkali metals, ammonium hydrosulphide, sulphur, phosphorus, and organic polymers [97]. However, the constituent that causes the variations in the color of Jupiter's clouds is unknown [98].

Simon-Miller et al., (2001) performed radiative transfer analysis using data taken by the *Galileo* spacecraft Solid State Imager (SSI) during its nominal mission (December 1995 to December 1997). The objective was to use the methane band at 727 nm and 889 nm, and color sensitivities at 410 and 756 nm to identify the vertical position of cloud absorption that leads to coloration. For this analysis, Simon-Miller et al., (2001) selected areas in the EZ, NEB, the GRS, a cyclonic and anti-cyclonic oval, and a bluish-gray, 5-μm equatorial hotspot.

It is believed that Jupiter's visible cloud structure is dominated by hazes in the stable stratified upper troposphere and stratosphere, and by condensate clouds of ammonia, ammonium hydrosulfide, and water at deeper levels (Fig. 4.2) [100]. West et al. (1986), have presented an interpretation of their observations. Historically, belts and zones have been thought to be regions of downwelling and upwelling, respectively, and therefore are expected to have cloud decks of differing densities and elevations (Fig. 4.3). This idea was supported somewhat by the visible differences in coloration, with belts being reddish and generally darker. Color differences could arise either because (a) the clouds in belts are older and covered by a reddening agent (e.g., photochemical smog that has rained down or sunlight processing of the cloud particles), with less overturning than is seen in the zones, or (b) because they are deeper in the atmosphere, allowing sublimation of the overlying ammonia ice rime on a core of redder material [103]. Neither of these speculations has yet been verified, because of a lack of detailed information about vertical motions and about the distribution of cloud and coloring with height and latitude [104].

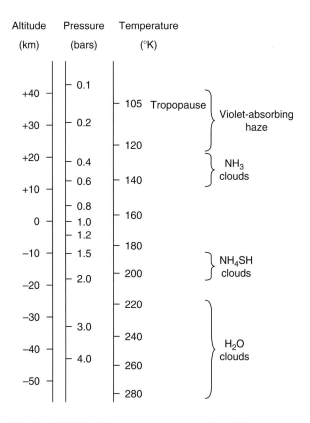

Fig. 4.2. A simplified model of cloud structure on Jupiter. (Adapted from Rogers 1995.) [99].

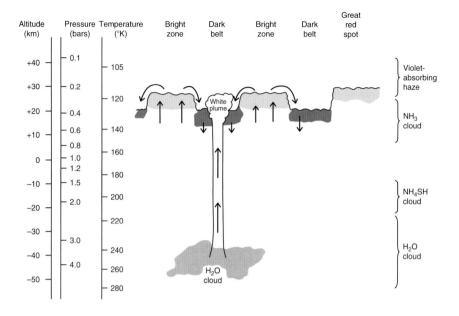

Fig 4.3. Model and cross section of Jupiter's cloud structure, showing direction of convection flow, height, pressure, and temperature. (Adapted from Rogers 1995 and Benton 2005.) [101, 102].

One theory is that hazes, or smog and aerosols in the stratosphere rain down and are deposited on the upper levels of clouds in the troposphere. We have smog on Earth, much of which is man-made. But, on Jupiter according to Amy Simon-Miller, "In Jupiter's case we likely have incident sunlight/ultraviolet radiation interacting with the aerosols and gases to make hydrocarbon smog. The small particles would then rain down toward the clouds. We are actually seeing the majority of the colored substance above the clouds in a thick hazy layer. The clouds underneath can be white as new ammonia ice wells up and the clouds overturn. In the locations where the clouds shoot higher from convection, they are whiter, since we are looking through less haze and seeing the fresh clouds." (Simon-Miller personal communication.)

According to Simon-Miller et al. (2001), the main differences between the belts and zones, if these areas are assumed to be typical, lies in the color of the tropospheric hazes and the optical depth of the variable cloud sheet. In all cases, the cloud sheet at the base of the tropospheric haze shows no evidence of coloration [105]. This generally agrees with the findings of West et al. (1896), and Simon-Miller et al. (2000), who find the majority of the coloration is likely to be in diffuse particles above the thick cloud deck. If white ammonia ice crystals are mixed in this layer, the color differences between belts and zones could be entirely due to the proportion of ammonia ice in the mix or riming the colored particles [106].

In the belts, the higher pressure of a thinner cloud sheet, and thicker, redder haze could be produced by downwelling and convergence in cyclonic sheer zones, evaporating the ices in the haze and clouds and increasing the layer thickness of tropospheric haze. A few fresh white clouds are forced up by convective events to replenish the layers as smog likely falls out from above to color them. In zones, small wisps of color could be formed in a similar manner, with weak cyclonic events temporarily evaporating the tropospheric hazes as they subside in the otherwise upwelling, and therefore brighter, surroundings [107].

In the case of the EZ, where a color gradient can sometimes be seen between the northern (EZn) and southern (EZs) sections of the zone (as mentioned previously), there may be an indication that a riming or mixing of ice with the haze caused the EZs to appear whiter than the EZn [108].

The Great Red Spot has its own internal energy driving it and the height and temperature arguments we use to explain color on the rest of the planet do not apply to it! (Simon-Miller, personal communication October 25, 2004). The Great Red Spot has been studied in detail with spacecraft imaging to determine its thermal structure, dynamics, color, and composition [109] (Fig. 4.4). Infrared data suggest that the GRS has a higher, colder cap of haze than other regions [110], and color studies indicate a difference in its color from other red regions [111]. Using data from the Galileo spacecraft SSIduring its nominal mission from December 1995 to December 1997, Simon-Miller et al. (2001), observed that the central core of the GRS was the reddest in color and may be caused by a different coloring agent than other regions on the planet [112]. Indeed, amateur ground based observations have revealed this same appearance. During the 2003–2004 apparition, the center of the GRS gave a distinct dark red condensation appearance at the core in many amateur CCD images. During the nominal mission, the cloud sheet was found to be quite thick at various locations within the GRS. The pressure results confirmed that the cloud sheet is higher (lower in pressure) in the north than in the south,

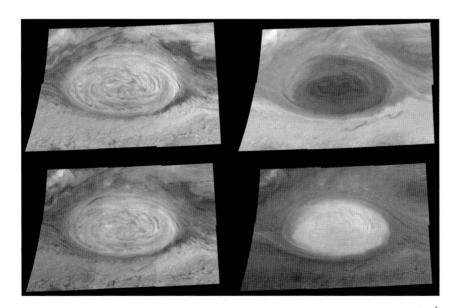

Fig. 4.4. Comparison of The Great Red Spot at Four different Wavelengths. These images show the appearance of the GRS in violet light (415 nm, *upper left*), infrared light (757 nm, *upper right*), and infrared light in both a weak (732 nm, *lower left*) and a strong (886 nm, *lower right*) methane absorption band. Reflected sunlight at each of these wavelengths penetrates to different depths and is scattered or absorbed by different atmospheric constituents before detection by the Galileo spacecraft. (Credit: Courtesy NASA/JPL-Caltech).

and is higher in the east than in the west, yielding an overall tilt as predicted. Other studies have previously implied that the GRS is tilted. Indeed, some have described the GRS as a tilted pancake! NW and SE regions of the GRS contained cloud sheets at about the same pressure, but differed substantially in haze structure. The tropospheric hazes for both bright and dark locations in the GRS were more-blue absorbing than those seen in the belt regions. The core locations also required redder stratospheric hazes, when modeled, than seen anywhere else on the planet, consistent with the possibility of a second differing coloring agent in the GRS [113].

It is common to see darker, bluish-gray features on the southern edge of the NEB at visible wavelengths (Fig. 4.5). For years we have referred to them as projections or the bases of festoons. It has been observed that the larger, dark patches are very warm at wavelengths of 5 μm, indicating that these are actually areas of thin clouds, allowing the warmth beneath to be detected in the infrared. These bluish-gray NEBs features are often referred to as 5-μm hotspots [114].

These dark bluish-gray features are regions of suspected strong downwelling, which clear out clouds and hazes allowing 5-μm radiation to escape from below [115] (Fig. 4.6). We see down into this hole, thus the darker appearance. David M. Harland states that an infrared hotspot is actually a hole in the ammonia clouds through which energy at a wavelength of 5 μm can readily escape. It appears dark at visual wavelengths because there is no ammonia to reflect sunlight [116]. Thus, it would appear that these dark areas, rather than being a feature, would instead be the absence of features, or clearings in the clouds. A hotspot appears bright in the

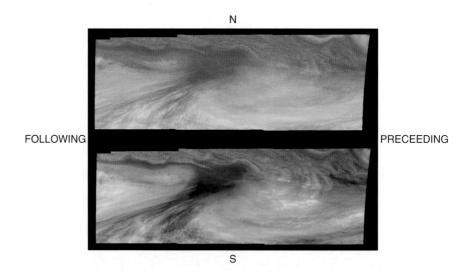

Fig. 4.5. A Galileo Spacecraft Image of an Equatorial Hotspot on Jupiter. These spacecraft images cover an area 34,000 km by 11,000 km. The top mosaic combines the violet and near-infrared continuum filter images to create an image similar to how Jupiter would appear to human eyes. The bottom mosaic used galileo's three near-infrared wavelengths (displayed in red, green, and blue) to show variations in cloud height and thickness. Bluish clouds are high and thin, reddish clouds are low, and white clouds are high and thick. The dark blue hotspot in the center is a hole in the deep cloud with an overlying thin haze. North is to the *top*. (Credit: Courtesy NASA/JPL-Caltech).

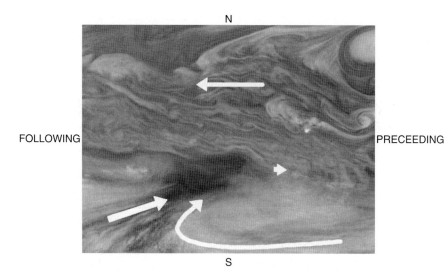

Fig. 4.6. Wind patterns in Jupiter's equatorial region and equatorial hotspot. Jupiter's atmospheric circulation is dominated by alternating jets of east/west (zonal) winds. The bands have different widths and wind speeds but have remained constant as long as telescopes and spacecraft have measured them. The top half of this Galileo spacecraft image lies within Jupiter's NEB, a westward (*left*) current. The bottom half shows part of the equatorial zone, a fast moving eastward current. The dark region near the center is an equatorial "hotspot". The clouds near the hotspot are the fastest moving features in the image, moving at ~100 m s^{-1}, or 224 miles per hour. The *arrows* show the winds measured by an observer moving eastward (*right*) at the speed of the hotspot, as though the hotspot is stationary while the rest of the planet moves around it. There is little cloud motion away from the hotspot, consistent with the idea that dry air is converging over this region and sinking, maintaining the cloud-free nature of the hotspot. North is at the *top*. (Credit: NASA/JPL-Caltech).

infrared because heat from the interior is leaking out through a hole in the upper layers of cloud [117]. These hotspots were a high priority for the Galileo spacecraft team [118].

The Galileo Probe descended through one of these "spots" and found little evidence of the ammonia cloud layer, only slight evidence for a cloud near the typical ammonium hydrosulfide cloud level, and no distinct water cloud [119]. Galileo data combined with that from the probe confirmed that these features mark where cold dry air is converging and being forced to descend. The hotspot examined by Galileo resided in an easterly jet stream. In effect, the prevailing eastward flow was pouring straight down the hole, maintaining its form [120] (Fig. 4.7). Galileo data indicated that the humidity in the vicinity of the hot spot through which the probe descended ranged from 0.02 to 10% with the lowest value in the center [121]. This explains why the probe detected less water molecules than expected. As the hot gas rises from the deep interior, the various volatiles precipitate out as rain. As the dry air "turns over" at the top of the atmosphere, the winds converge and descend. By this point, there are no volatiles left to condense to form clouds, so a dry clearing is created. Then, as the cold air descends, the pressure rises and it is heated again. According to Glenn Orton, "these dry spots may grow and diminish, but they return in the same places, possibly because of the circulation patterns" [122].

Fig. 4.7. Three dimensional visualization of Jupiter's equatorial region. An equatorial "hotspot", a hole in the bright, reflective, equatorial cloud layer where warmer thermal emission from Jupiter's deep atmosphere can pass through. Dry air may be converging and sinking over these regions. The bright clouds to the right of the hotspot may be examples of upwelling moist air and condensation. This is a view from between cloud layers and above the streaks in the lower cloud leading toward the hotspot. The cloud streaks end near the hotspot, consistent with the idea that clouds traveling along these streaks descend and evaporate as they approach the hotspot. View is to northeast. (Credit: Courtesy NASA/JPL-Caltech).

Thus, this hotspot that the probe descended into was a relatively clear and dry area. Eight to ten somewhat evenly spaced hotspots are present at this latitude at any given time; located at the boundary between the northern part of the equatorial zone and the southern edge of the northern equatorial belt, between 6° and 8° N. All the hotspots move as a group with a velocity near 103 m per second with respect to System III, at the time of the probe entry. The hotspots are also associated with accompanying equatorial plumes; optically thick cloud features which are believed to be regions of strong convection [123].

The 5-μm hotspots lie in the jet at the boundary between the equatorial zone and the north equatorial belt (NEB), a shear zone believed to be a region of general downwelling air [124]. Not all bluish-gray features are hotspots, but all hot spots are bluish-gray features.

Where the probe entered Jupiter's atmosphere there was no evidence of a deep-water cloud. The favored explanation is that the probe entry hot spot was an extremely dry downdraft [125]. The analysis of Orton et al., of the vertical structure of hotspots is that there is a two-layer cloud structure (as opposed to the normal three-layer one). There is, (1) an upper tropospheric cloud layer above 450 mbar probably consisting of ammonia ice particles less than 1-μm in size with significant opacity in the visible, becoming optically thin in the near-infrared and negligible at mid-infrared wavelengths. (2) A tropospheric cloud below the 1 bar level, probably ammonia hydrosulphide, with small optical depth, ≤ 1.0 at 4.78 μm.

The normal third layer of distinct water clouds is absent. According to Orton et al., comparisons with regions to the north and south of hotspots are consistent with the interpretation that hotspots are regions of reduced cloud opacity, possibly because of a dry downdraft. A thicker cloud of possibly brighter particles in the upper troposphere is required to satisfy the highly reflective cloud features in the visible, as well as particles of size greater than ~3 μm to satisfy the opacity in the mid-infrared. In other words, the high, bright cloud material that must be surrounding the hotspot is absent from within the hotspot itself. Thus, the hotspot is a region of reduced cloud opacity, because of the dry downdraft [126]. Thus, an infrared, or 5 μm, hot spot is actually a hole in the ammonia clouds through which energy at 5 μm can readily escape. It appears dark at visual wavelengths because there is no ammonia to reflect sunlight. We can assume that, had the probe descended into any other region of Jupiter's atmosphere, it would have detected more water, and that the 'typical' atmosphere of Jupiter is, in fact, wetter.

Five micrometer hotspots have a lifecycle. The Galileo probe entered Jupiter's atmosphere on December 7, 1995. Hotspots were being monitored preceding the arrival of the probe, and the expected probe entry site (PES) had been determined long before. During the months preceding the probe's arrival, the morphology of this PES was quite interesting. In September 1995 the PES hotspot apparently had merged with another hotspot and then split again soon after. At the beginning of what Orton et al., referred to as what might be called the beginning of its 'life cycle', between October 3 and 13, 1995, the hotspot reemerged as a small 'wedge shape' evolving into a larger 'comma' shape, with a bright but small round core and a small 'tail'. Then a transient filament-like morphology appeared, after which it evolved into a 'mature phase', when the spot covered an area a few degrees in longitude and had a tail tilted approximately 30°. Subsequently, it reached its most intense state, flattened and expanded enormously. The PES remained in this flat morphology until between November 1995 and July 1996. Following this, it began its life cycle again assuming a small wedge shape, followed by a comma shape. It skipped the mature phase then began again as a small feature in December 1996. By August 1997 it had evolved into a mature phase, and began once again to break up again shortly thereafter [127].

A study spanning more than three years was made of the longitudinal locations, morphology, and evolution of 5-μm hotspots using the Infrared Telescope Facility-National Science Foundation Camera (IRTF-NSFCAM). This three-year period included the date of the Galileo probe entry. According to Orton et al., an analysis of the data shows that within periods of several months to a year, there are eight or nine longitudinal areas with high likelihood of containing a 5-μm hotspot. These areas drift together (approximately at the same rate) with respect to System III at a rate that changes only slowly in time, and they are quasi-evenly spaced, suggesting a wave feature [128].

Ortiz et al., observed that the number of hotspots is actually almost always higher than 10 or 11. They also observed that there might be different wave modes, or wave numbers, which move at slightly different speeds. This could explain why individual hotspots seem to move a little faster than others. Although their morphology is complex, most hotspots seem to show a mature phase in which they are larger, with a hot narrow festoon extending south and westward from the eastern most edge, tilted about 30°. More observations are needed to provide a complete correlation, if one exists, between the 5-μm hotspots and the bluish-gray festoons we see visually [129].

It had been thought that these features formed at random longitudes around the planet, but a closer inspection of the data revealed that this was not true once an appropriate drift rate was chosen. Ortiz et al., concluded that both the pattern and observed speeds were consistent with a Rossby wave [130].

Ortiz, define a hotspot as a region in Jupiter's atmosphere whose equivalent brightness temperature at 4.8 μm is greater than 240 K at nadir viewing. The hotspots extend from the southern edge of the NEBs and into the equatorial zone (EZ). They are not referred to as NEB or EZ hotspots, but by their central latitude, that is 6.5° N planetocentric [131].

High-resolution red and near-infrared images reveal that regions of exactly the same morphology as the hotspots at 4.8 μm are very dark visually. However, not all the dark features seen in the red and near infrared are bright at 4.8 μm. Thus, while every hotspot is associated with a bluish-gray feature, not every bluish-gray feature (festoon) is associated with a hotspot [132]. According to Orton, all of the bluish-gray features are warm, even if slightly so, in the infrared; however, as Ortiz writes, not every bluish-gray feature would be classified as a hotspot (Orton, personal communication, August 2005).

Long-term observations of the NEBs bluish-gray features (the visible festoons) reveal that their drift rates and locations show similarities to the 5-μm hotspots; that is, their quasi-periodic, but often symmetric, spacing in longitude as well as their time variable numbers around the full circumference of the planet. According to Rogers (1995), and the data of other organizations, these 'dark NEBs projections' often have lifetimes of months with a faded feature reappearing in the same location [133].

We have previously discussed the physical appearance of these features (festoons) as seen visually in amateur instruments and how during one period the A.L.P.O. made a special effort to follow the festoons from one apparition into the next. I think it would be a wonderful project for amateurs to repeat that endeavor on a more continued basis with the goal of tracking the morphology just described.

These bluish-gray features appear "bright" at 5 μm, that is to say, in the infrared. Since infrared senses differences in temperature, we know they are hot spots. Since Earth's atmosphere shields us from infrared wavelengths, work in the infrared must be done at high altitude. Orton has gathered his data with the NASA infrared telescope on Mauna Kea, Hawaii. While infrared work is still much the domain of the professional astronomer, how far off in the future can amateur instrumentation and capabilities be?

While we often see extended features in Jupiter's belts and zones that are somewhat dark, true condensed, independent spots that are truly dark are rare in Jupiter's atmosphere. We have already had some discussion of the South Temperate Dark Spot of 1998. It was one of the darkest spots ever seen (Fig. 4.8). The Galileo spacecraft was used to examine this spot. Data indicated that the spot was warmer than the environment in which it resided. The warm cloud-free conditions indicated it was a region where dry upper atmosphere gas flow had converged, made a hole in the cloud as it was forced to descend, and warmed as its density increased [134].

The 1998 Dark Spot (South Temperate Dark Spot of 1998) was unlike the cloud-free hotspot into which the probe had descended. According to Dr. Glenn Orton, JPL, both appear warm at 5 μm, but the dark spot was truly warmer than its surroundings, whereas the '5-μm hotspots' actually appear to be the same temperature as their surroundings [135].

There is also a distinct color contrast. The 5-μm hotspots, festoons as we see them, are normally colored a dark blue-gray. The 1998 Dark Spot was spectacularly

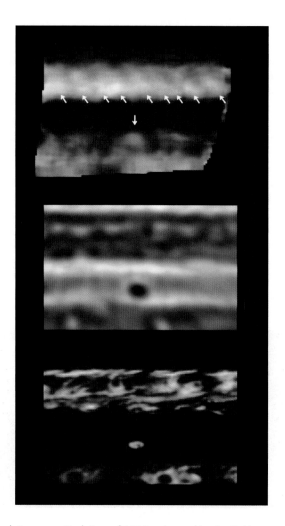

Fig. 4.8. The South Temperate Dark Spot of 1998 as Imaged by the Galileo spacecraft. The single arrow in the upper image, which is a map of Jovian temperatures, identifies a warm area that is further identified with the "black spot", shown in the visible light image of the *middle* image. The spot may be the result of a downward spiraling wind that blows away high clouds and reveals deeper, very dark cloud layers. The bottom image is a thermal radiation image sensitive to cloud-top temperatures. The warm temperatures and cloud-free conditions imply that this feature is a region where dry upper-atmospheric gas is being forced to converge, is warmed up and then forced to descend, clearing out clouds. (Credit: NASA/JPL-Caltech).

black. Visually when I observed the spot, it was incredibly small, yet almost as black as a moon shadow.

In late 2005, the remaining white south temperate oval BA changed color and became noticeably red in early 2006. I think this was quite unexpected by the astronomical community, and proved to be of great interest to professional astronomers. Oval BA is the remnant of three long-lived white ovals that formed in the 1930s south of the GRS. The previous two surviving ovals, BE and FA collided and merged in April 2000 (Fig. 4.9). Oval BA had always displayed an off-white or dusky-white appearance, but images obtained by amateur astronomers in late 2005 caught the oval displaying a tawny appearance. Images taken in December 2005 revealed that oval BA was red for the first time in its history. This was confirmed by further imaging in early 2006 [136] (Fig. 4.10).

According to Simon-Miller et al., Hubble Space Telescope (HST) images from 1998 to 1996, and additional Galileo spacecraft data from 1997, indicated that oval BA was similar to the rest of Jupiter's white zone regions with high, optically thick, white clouds and haze. In contrast, the darker belt regions appeared to be comprised of deeper white clouds covered with a thick, blue-wavelength, absorbing (or scattering) haze. This haze is actually widespread, but appears darkest in locations with the deepest clouds, allowing for the longest path length through the haze. As previously mentioned, the GRS with its high, thick, colored clouds differs from Jupiter's other anticyclonic regions and ovals. A separate coloring agent is required to explain its color. This coloring agent may be material that is dredged up from deep within the atmosphere in the upwelling central region of the GRS' strong, high-pressure system. Neither the coloring agents in the belts nor the

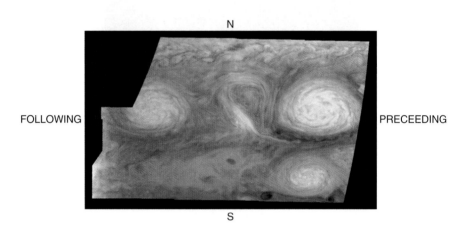

N

FOLLOWING

PRECEEDING

S

Fig. 4.9. Jupiter's Long-Lived South Temperate White Ovals in False Color. Anticyclonic (counterclockwise) ovals BE and FA as imaged by the Galileo spacecraft. Note the cyclonic (clockwise) intervening balloon shaped vortex between them. When this smaller oval disappeared the two larger ones merged. In this false-color image, light blue clouds are high and thin, reddish clouds are deep, and white clouds are high and thick. The clouds and haze over the large white ovals BE and FA are high, extending into Jupiter's stratosphere. North is at the *top*. (Credit: NASA/JPL-Caltech).

Fig. 4.10. South Temperate Oval BA in Conjunction with the Great Red Spot. The Great Red Spot on July 15, 2006 with a dark center. Note the collar of gray material bordering the GRs on its southern edge. South Temperate Oval BA, just to the south, has also taken on a reddish color, very similar in intensity to the GRS. (Credit: Courtesy Donald C. Parker).

suspected second coloring agent in the GRS has ever been conclusively identified. If there is a second coloring agent in the GRS, it is unclear whether it is produced at some depth or if it is caused by the UV irradiation of lofted material over the GRS. Since cyclonic systems can also be red, it would appear that something from depth should be responsible or at least a contributor to the reddening. This material from depth might be always red or it may turn red over short exposure to UV photons [137]. The compound NH_4SH could be a possibility, since it turns red in only a few hours when exposed to UV photons (Simon-Miller, personal communication).

Images taken by HST on April 8, April 16, and April 24, 2006 indicated that oval BA was now clearly red. However, white clouds seen in the methane-band images indicated that the color was not caused by a longer path length through colored haze, nor was it a clearing out of haze and clouds. This was further evidence of a second coloring agent [138]. There have been other, short-lived, storms that are cyclonic that have taken on red coloration. When they are super red, they seem to be free of any overlying haze, indicating big holes in the upper hazes. Again, this is a good argument that there is something from below that can be quite red (Simon-Miller, personal communication). The presence of a second coloring agent in both oval BA and the GRS indicated that the storm had likely intensified and was isolating material in much the same manner as the GRS. A comparison of rotational velocities measured by the Voyager and Galileo spacecraft, and later by the Cassini spacecraft and the HST shows increased vorticity. This increased vorticity could provide the mechanism for dredging material from some depth, allowing the darkening of oval BA. Thus, this normally white storm appears to have intensified to become more like the GRS [139]. Future observations of oval BA may lead to clues that will finally unlock the mystery into the red color of the GRS and other ovals. Certainly, amateurs with CCD and web cams will play an important role in this research.

There have been entire studies devoted to determining the cause of Jupiter's color, and there are theories that have gained support among scientists. However, most are unproven. While we are fairly sure that the white clouds of the zones are made of ammonia ice [140], the most fundamental questions about Jupiter, such as what colors its clouds, have yet to be answered [141].

In summary, we believe that color on Jupiter does not depend upon chemical composition. Potential color substances, agents that affect the shape of the reflection spectrum, are referred to as chromophores. Astronomers believe that Jupiter's visible cloud structure is dominated by hazes in the upper troposphere and the stratosphere, and by condensate clouds of ammonia, ammonium hydrosulfide, and water at deeper levels. The hazes in the upper troposphere and stratosphere may interact with ultraviolet light, producing hydrocarbon smog that rains down, coloring the cloud tops in the troposphere. The distinct clouds in the troposphere that reside at different depths, may also affect the visual appearance of color. Ammonia ice clouds that well up to higher altitudes due to convection will appear white. The core of the GRS is the reddest of all areas on the planet. A coloring agent, different from the rest of the planet, may be responsible for the redness of the GRS. And the bluish-gray 5-μm hot-spots of the NEBs are not colored at all, but result from the absence of ammonia clouds.

Hopefully the preceding discussion, though somewhat complicated, will help us understand the difficulties involved in understanding color in Jupiter's atmosphere. And, while the techniques discussed at present might be the realm of the professional, it should be noted that many amateurs today are imaging Jupiter in methane and other wavelengths. Imaging in different wavelengths provides valuable clues to the height and structure of the clouds. And who knows, with the advances being made in instrumentation available to amateurs today, more and more areas of study will cease to be the sole realm of the professional. Some understanding of these disciplines now can prepare amateurs for their participation in the future.

4.2 The Chemical Composition

As mentioned previously, spectra of Jupiter's clouds have not provided many answers regarding the chemical composition of the planet, and all of the major compounds predicted to form clouds are chemically simple and would appear white. However, because of Jupiter's size and mass, it is expected to have retained chemical elements similar to the cosmic proportions when it formed – that is, the proportions found in the Sun and the interstellar gas [142]. Therefore, as we understand the formation of our solar system, we would expect hydrogen and helium to make up the greatest part of Jupiter's mass, or 99%. The hydrogen-dominated atmosphere is chemically 'reducing' – that is, the other elements should exist mostly as hydrogen compounds [143]. Over the years ground based, and more recently, spacecraft observations have contributed significantly to our knowledge of Jupiter's chemistry. We know that different elements exist in Jupiter's clouds at different levels or bars of pressure, since their formation could only occur at specific pressures. Therefore, as different wavebands are emitted from different depths, a spectral line reveals the concentration of a gas at and above the corresponding depth at that wavelength [144]. We shall see later how this property helps to characterize the vertical structure of Jupiter's atmosphere.

The first molecules that were discovered on Jupiter were methane and ammonia. Rupert Wildt (1931, 1932) identified them from several absorption bands that had long been known in the red-light spectrum [145]. After the Voyager spacecraft missions, and prior to the Galileo era, we knew that the following molecules were present in Jupiter's atmosphere [146].

Methane (CH_4) is the most abundant gas after hydrogen and helium. It exists at all levels of the atmosphere [147].

Ammonia (NH_3) is the next most abundant, existing at 1 bar or deeper. Ammonia levels seem to be affected by Jupiter's weather. For example, there is less ammonia in the belts than in the zones. This pattern is consistent with the theory that belts are regions of sinking gas or downwelling, and the ammonia poor gas drawn down from higher altitudes explains why the ammonia clouds are lower and thinner in the belts [148].

Water (H_2O) is much less abundant than ammonia. There is none at 1 Bar and only a little at ~4 Bars [149].

Hydrogen sulfide (H_2S) has not been definitely detected, but it is expected to be present deeper, below the clouds [150].

Phosphine (PH_3), *germane* (GeH_4), and *arsine* (AsH_3) (Noll and colleagues, 1990), have been detected in proportions similar to those which are predicted from cosmic abundances. Since they should be unstable in the troposphere, their presence may be due to updrafts from much deeper levels. However, so far there is no definite sign of variations in their abundances between the belts and zones [151].

Carbon monoxide (CO) is even more surprising as it is an oxidized compound that must be unstable in reducing atmospheres. However, the observations show it to be well mixed throughout the troposphere. Therefore, it must also be brought up from deeper regions [152]. According to Lewis and Fegley (1984), the abundances of phosphene, germane, and carbon monoxide can all be explained by updrafts from the level of Jupiter's atmosphere where the temperature is ~800–1,300 K, a depth well below any direct probing [153].

Hydrogen cyanide (HCN) was first definitely detected by Tokunaga and colleagues (1981), and is not in chemical equilibrium. Its origin is uncertain, its most likely source being photochemical (light-induced) reactions in the upper troposphere [154].

Acetylene (C_2H_2) and ethane (C_2H_6) have also been detected, but in infrared emission, indicating they are in the stratosphere. It is believed that acetylene and ethane are produced photo-chemically by irradiation of methane in the stratosphere [155].

Since the Voyager missions, the Galileo spacecraft has orbited Jupiter and the Cassini spacecraft has made its historic flyby on its way out to Saturn. As a result of these two missions, we now have evidence of additional molecules. We also know that the origins of these gas molecules are linked to internal thermo-chemistry, photo-chemistry, and impactor chemistry; and these processes are now more fully understood with respect to Jupiter.

According to Kunde et al., "down in the deep atmosphere where pressures and temperatures are high, Jupiter's thermo-chemical furnace processes the approximately solar elemental composition by converting H (hydrogen) atoms into molecular forms (H_2) and reactive atoms (e.g., C, N, and O [carbon, nitrogen, and oxygen]) into saturated hydrides (methane, CH_4; ammonia, NH_3; water, H_2O). Convection transports these molecules upward into the cooler regions, where H_2O, NH_4SH, and NH_3 condense to form clouds" [156]. Thus, Jupiter is a very good example of thermo-chemical processes.

Photo-chemistry relates to the changing of molecules due to the interaction of UV photons with molecules in Jupiter's upper atmosphere, or bombardment of the surfaces of asteroids and meteors by solar UV.

The composite infrared spectrometer (CIRS) on board the Cassini spacecraft also detected two new hydrocarbon species in Jupiter's stratosphere, the methyl radical CH_3 and diacetylene C_4H_2. Both of these elements contribute to Jupiter's stratospheric photo-chemistry. They were detected in Jupiter's north and south auroral infrared hotspots [157]. The polar auroral stratosphere is driven by the deposit of energetic magnetospheric electrons and ions that heats the atmosphere, enhances the abundances of some hydrocarbons by ion-induced chemistry, and increases the visibility of all stratospheric species in the thermal infrared by elevating the ambient temperatures. CIRS measurements of Jovian auroral regions show that the emissions of many hydrocarbons within the auroral infrared hotspots are enhanced compared to the surrounding ambient polar atmosphere; that is, there is a distinct difference in temperature and/or composition between the hotspot relative to its surroundings [158].

Impactor chemistry relates to the influence of chemical composition by external sources of material. The most dramatic illustration of this was the multiple impacts of comet Shoemaker-Levy 9 (SL9) into Jupiter in July 1994. CIRS observed spatial distributions of carbon dioxide and hydrogen cyanide and both are considered to be by products of the SL9 impacts. SL9 injected large quantities of nitrogen, oxygen, and sulfur bearing molecules into Jupiter's stratosphere. Substantial amounts of hydrogen cyanide, carbon monoxide, and carbon monosulfide were produced in the resultant shock chemistry and subsequent photo-chemistry. Carbon dioxide was then produced from the photo-chemical evolution of CO and H_2O [159].

The probe from the Galileo spacecraft made in situ measurements as it descended into Jupiter's cloud tops. The probe helped to determine that the helium abundance in Jupiter's atmosphere was in fact very close to what was expected, just under 25% of the solar value. By contrast, the abundances of methane, ammonia, and sulfur exceeded the solar abundances, which implies that the infall of small bodies such as comets into Jupiter has played an important roll in the planet's evolution. Also, argon, krypton, and xenon were found to be in much greater abundance than solar values, two to three times what was expected, meaning that Jupiter did not form solely from the solar nebula [160].

Now, with our more complete understanding of Jupiter's chemical composition, we know of these additional molecules in Jupiter's atmosphere: *Hydrogen (H_2), Methyl radical (CH_3), Ethylene (C_2H_4), Methylacetylene (C_3H_4), Benzene (C_6H_6), Diacetylene (C_4H_2), Carbon dioxide (CO_2),* and the isotopes *Deuterated hydrogen (HD), Monodeuterated methane (CH_3D), Isotopic methane ($^{13}CH_4$), Isotopic ethane ($^{13}C_2H_6$), Isotopic ammonia ($^{15}NH_3$),* and tropospheric ices *Water ice (H_2O ^{Ice}), and Ammonia ice (NH_3 ^{Ice})* [161] (Kunde et al., supporting online information, pp 12–13, Science 2004). Also, *argon, krypton,* and *xenon* [162].

4.3 The Vertical Structure of Jupiter's Atmosphere

Unlike the physical or visual appearance of Jupiter where observations of changes in the longitudinal and latitudinal positions of features has long been an area in which amateurs have been able to participate, observations of Jupiter's vertical structure, at least at the time of this writing, have not. Indeed, even the ability of the professional astronomer to pierce deeply the cloud tops of Jupiter is still severely restricted, as we shall see. However, as amateurs we should seek to understand that knowledge which is available and the processes by which such data is acquired and reduced.

With an understanding of the abundances of elements in Jupiter's atmosphere, and knowing the profile of temperature and pressure with altitude, it is possible to predict the levels at which clouds of various types should form [163]. The depth to which a sensing instrument can view depends upon the opacity of the gas, which is dependent upon the chemical composition. Different gases condense at different temperatures and pressures, both of which increase with depth [164]. Figure 4.1 depicts a simplified model of the vertical structure of Jupiter's atmosphere.

John S. Lewis (1969) first developed the cloud model presently accepted by Jupiter scientists. Lewis predicted three main cloud layers: ammonia ices at the highest level (0.3–0.7 bars), ammonium hydrosulphide below it (NH4SH; 2 bars), and water deeper still (5–6 bars) [165]. Water will condense as it rises within Jupiter's atmosphere and passes through the 4-bar pressure zone. Ammonia condenses at much colder temperatures, and higher in the atmosphere at ~0.7 bars. But, since Jupiter's atmosphere is not cold enough to condense methane, it remains in the gaseous state [166]. The water clouds could include an upper layer of ice crystals and a lower layer of water droplets, like on Earth. The water is not pure, but would have a lot of ammonia dissolved into it (Fig. 4.2). According to Carlson and colleagues (1988), the ammonia clouds are like thick cirrus and are unlikely to precipitate anything. However, the thick water–ammonia clouds below them will probably produce rain [168].

Because the methane is well mixed, and because it has convenient spectral absorption features, it can be used as a 'tracer' for determining the vertical distribution of clouds of other elements. This is why professionals today are so excited that amateurs are now able to image Jupiter in methane wavelengths. The atmospheric pressure of a cloud can be inferred from images taken in the near infrared. Knowing the pressure of a cloud allows us to infer its depth, and knowing its depth allows us to infer its chemical composition. This is because clouds of various types will only form at specific pressures and temperatures. Or, considering this from another point of view, if an observation identifies a cloud of ammonia on Jupiter for example, then we would know the depth of the cloud and its temperature. An image taken at a wavelength which is strongly absorbed by methane can penetrate no deeper than 1 bar; a wavelength which is less strongly absorbed can penetrate a little deeper; and a wavelength which methane does not absorb will penetrate to about 8 bars. If we make assumptions about opacity, then such imagery can be processed to infer the pressure, and thus the depth, at which a feature resides. By 'stacking' the three images it is possible to infer the vertical structure of a specific cloud [169]. This is very useful information.

With a simplified model we can portray the vertical cloud structure of Jupiter's belts and zones, and the convective flow (Fig. 4.3). Note the relative altitudes of the clouds and hazes in the belts and zones. Also note that the top of the GRS resides at the highest altitude.

It may surprise those new to observing Jupiter that we are not able to penetrate very deep into Jupiter's atmosphere with our instruments. Remember, the clouds of the troposphere are optically thick, and we only see the tops of clouds illuminated by sunlight. The stratosphere has lower densities and is mainly gases and aerosols, and is optically thin. In fact, the stratosphere is invisible visually. According to Simon-Miller, the stratospheric haze is also virtually impossible to see, except in the UV, where it is usually featureless, though some features are occasionally seen. (Simon-Miller personal communication). Certainly, getting down to 4 or 5 bars is just scratching the surface. So, what might exist deeper in the planet?

To infer the structure of Jupiter far below the clouds, we must turn to theoretical physics, using what information is available on the behavior of hydrogen and helium at extreme temperatures and pressures [170]. The current accepted theories and models seem to agree on the following main features (Fig. 4.11). Below the visible clouds exists a deep atmosphere/ocean of molecular hydrogen. At the higher altitudes (lower pressure) the hydrogen exists as a gas, but lower down, ~1,000 km below the cloud tops at thousands of bars of pressure, this hydrogen gas turns into a hot liquid [171]. Even further down, at a depth of 15,000–25,000 km below the cloud tops at temperatures over 15,000 K, at a pressure of 2–4 million bars we should find a mantle of metallic hydrogen [172]. And finally, at the very center of the planet, at a pressure approaching 1,006 million bars and a temperature over 35,000 K, there should exist a rocky core. At this pressure and temperature, most of the planet's metals should have sunk into this core [173].

Jupiter's interior generates extreme heat. What causes this tremendous heat? Earth's weather is driven by heat absorbed from the Sun. However, Jupiter's weather is driven more by its own internal heat than from solar heat. Spacecraft

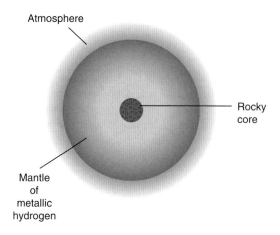

Atmosphere

Rocky
core

Mantle
of
metallic
hydrogen

Fig. 4.11. A simplified model of Jupiter's interior. (Courtesy John W. McAnally).

observations have determined that Jupiter actually emits 1.67 times more heat than it absorbs from the Sun [174].

Jupiter radiates the heat from its very hot interior. The prime candidate for this heat is primordial and gravitational heat since the planet has not finished cooling and contracting since it formed. Or, alternatively the heat could be caused by contraction of the planet at 1 mm per year [175].

Although amateurs cannot directly observe the chemistry, pressure, and temperatures that drive the activity in Jupiter's atmosphere, a basic understanding of these mechanisms can help us appreciate even more the visible features we observe. By understanding the processes that create the features we see, we gain an insight into the planet not perceived by visual observations alone. To me, every aspect of the study of Jupiter is fascinating. But, the understanding of pressure and temperature and how they affect cloud structure has new meaning for amateurs. We have discussed how methane, ammonia, and other elements exist at various pressures and temperature, and how observing at specific wavelengths can provide information concerning the vertical structure of the clouds. Today, the ability of amateurs to observe at anything other than visible and methane wavelengths is limited. However, advances are being made. In recent years, more than just a handful of amateurs are observing in the methane band. Recently when I sent an amateur's CCD image of Jupiter in methane to Dr. Reta Beebe at New Mexico State University, she remarked to me, "Now, this is very useful!" (Beebe, personal communication). Methane is only the beginning. Thus, an understanding of Jupiter's atmospheric structure becomes even more important. Surely in this century, with technology advancing so rapidly, amateurs will find more and more wavelengths at their disposal. I think before the century is over, amateurs will be doing what professionals are doing today.

Chapter 5

The Electromagnetic Environment Surrounding Jupiter

There is more to Jupiter than meets the eye, at least the amateur's eye in the visible spectrum, such as its electromagnetic environment. Jupiter's electromagnetic environment is a fascinating aspect of the planet and a basic understanding of this environment will make you a more knowledgeable amateur astronomer.

5.1 The Magnetosphere and Magnetic Field

Jupiter is encased in its own magnetosphere as Earth is, with one big exception. Jupiter's magnetosphere is big, really big. According to Harland, "Jupiter's magnetosphere is the largest discreet structure in the solar system. It is millions of kilometers across and tens of millions of kilometers long. In fact, if it could be seen by the naked eye, it would appear larger than the Moon in our night sky – it would span 1.5°, compared to 0.5° for the Moon" [176]. Hill describes it as approximately 20 solar diameters wide and several astronomical units in length (Fig. 5.1) [177]. Now that is an enormous structure!

In attempting to comprehend the great size of Jupiter's magnetosphere, we can express it as a multiple of Jupiter's radius. Jupiter's radius is measured from its center to its "surface"; R_J represents one Jupiter radii, or 71,400 km. The distance from Jupiter's center to the bow shock can be $100 R_J$. Jupiter's magnetotail can stretch to a radius of $150–200 R_J$. Jupiter's magnetosphere is so large because it is inflated by hot plasma and by the centrifugal force of the cooler co-rotating plasma sheet, to be discussed.

The magnetosphere is divided into three zones (Fig. 5.2). The outer zone is the most rarified and its shape and structure are variable. The middle zone contains an equatorial sheet of plasma that co-rotates with the planet's magnetic field. The inner zone contains the Io torus, the densest part of the magnetosphere. The structure of the magnetosphere can be further described as consisting of several parts. Its outer boundary is formed by its interaction with the solar wind – that tenuous magnetic field and plasma that streams outward from the Sun. The solar wind wraps itself around the sunward side of Jupiter's magnetosphere, spilling around the planet as

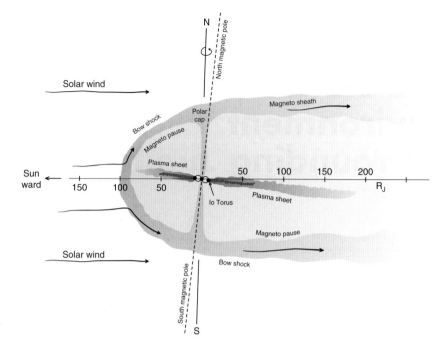

Fig. 5.1. A model of Jupiter's magnetosphere (After Rogers 1995 [178]).

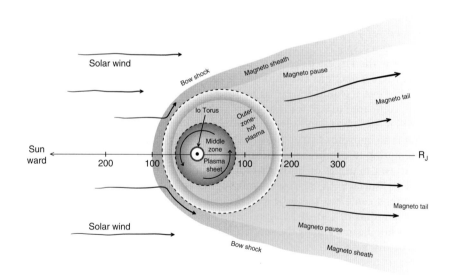

Fig. 5.2. A model of Jupiter's magnetosphere – looking down from above the Jovian north pole. (After Rogers 1995 [179]).

it is deflected. This sharp boundary is called the 'magnetopause'. Plasma in the solar wind is sharply deflected as it approaches the magnetopause, creating a 'bow shock'. The area between the bow shock and the magnetopause is called the 'magnetosheath' [180]. We know that the distance of these boundaries from Jupiter varies with the strength of the solar wind, which also varies over time [181].

The pressure balance at the magnetopause determines the size of a magnetosphere. The internal pressure is sustained by the planet's magnetic field and the plasma (hot ionized gas) trapped inside the field. The external pressure is provided by the solar wind – a fully ionized stream of plasma that flows continually, but variably, outwards from the Sun. The wind is highly supersonic, and its pressure variations tend to steepen into shock waves as they move away from the Sun. These interplanetary shocks produce sudden compressions and expansions of magnetospheres [182].

In 1995 as the Galileo spacecraft approached Jupiter, scientists eagerly awaited the detection of the magnetosphere by the spacecraft. The magnetosphere was finally detected by the spacecraft's magnetometer when a lull in the solar wind allowed Jupiter's magnetosphere to balloon outward. The bow shock of the magnetosphere actually washed back and forth over the spacecraft several times between November 16 and November 26, 1995 [183]. Thus, this single-spacecraft event demonstrates that Jupiter's magnetopause moves inward or outward in response to the pressure of the solar wind.

At Earth, multiple-spacecraft observations have been used to understand how this motion occurs, since this type of observation can give us a snapshot view of a transient event in progress [184]. According to Fairfield and Sibeck et al., variations in the size of Earth's magnetosphere have been reported for decades, some 1,821 observations [185].

At Jupiter, fewer but similar observations have been made with Pioneer and Voyager data, and it is well recognized that the magnetopause distance should be controlled, at least in part, by the pressure of the solar wind [186].

The Cassini and Galileo spacecraft obtained simultaneous observations during December 2000 and January 2001. The Cassini spacecraft was on its way to Saturn, and the Galileo spacecraft was in orbit around Jupiter at the same time. Scientists took advantage of the Cassini flyby, centered on December 30, 2000, while Galileo was still in its extended orbital mission around Jupiter. As Cassini approached the vicinity of Jupiter, it first encountered Jupiter's bow shock on December 28, 2000. Cassini would actually encounter the bow shock numerous times. The data indicate that Cassini actually encounters Jupiter's bow shock at least 16 times between December 28, 2000, and January 20, 2001, as the pressure of the solar wind increased and the bow shock washed back and forth over the spacecraft [187]. During this encounter with Jupiter, Galileo was well sunward of Cassini. An analysis of the data taken by the two spacecraft indicate that the magnetopause was in a state of transition from a significantly inflated size at Cassini to a large but nominal size at Galileo. Based on the distance between the two spacecraft and the solar wind pressure, it would take some 5 hours for the pressure front to propagate from the position of Galileo to that of Cassini. Hence, there was clear evidence of a region of increasing pressure moving from Galileo to Cassini's position on timescales similar to what would be expected for a region of increased pressure in the solar wind [188]. Thus, we see that Jupiter's magnetosphere is a constantly changing structure on a scale and timetable we can scarcely imagine, due to the influences of the solar wind.

Although a vacuum by our everyday standards, Jupiter's magnetosphere contains rarified plasma – that is, gas dissociated into positively charged ions and negatively charged electrons. Even larger than the magnetosphere is a rarefied nebula of sodium atoms that envelops the entire Jovian system. Several spacecraft missions have reported the presence of large numbers of energetic ions and electrons surrounding Jupiter. Apparently, relativistic electrons are detectable for several astronomical units from the planet. Energetic neutral particles have also been detected, neither the mass nor charge state of which can be determined, and so are labeled energetic neutral atoms. Images have also shown the presence of sodium as a trace element. The Cassini spacecraft discovered a fast and hot atmospheric neutral wind extending more than 0.5 astronomical units from Jupiter, and the presence of energetic neutral atoms accelerated by the electric field in the solar wind. It is thought that these atoms originate in volcanic gases from the moon Io, and undergo significant changes through a number of electromagnetic interactions, then escape Jupiter's magnetosphere into the Jovian environment. This 'nebula' extends outward from Jupiter over hundreds of Jovian radii [189].

We are going to use the term 'plasma' a lot, so we should understand what plasma is. *Plasma* is highly ionized gas, consisting of almost equal numbers of free electrons and positive ions. The *plasma sheet* is low-energy plasma, largely concentrated within a few planetary radii of the equatorial plane, distributed throughout the magnetosphere throughout which concentrated electric currents flow [190]. Because it is made up of low-density ionized gas, plasma is a very good electrical conductor with properties that are strongly affected by electric and magnetic fields. Individual ions and electrons interact with one another by both emission and absorption of low-frequency 'waves'. Plasma waves occur both as electrostatic oscillations - which are similar to sound waves – and as electromagnetic waves. Such waves are induced by instabilities within the plasma [191].

The magnetosphere owes its existence to Jupiter's magnetic field. Like that of Earth, Jupiter's magnetosphere consists of plasma populations that are mostly confined to certain regions of the magnetic field and are mostly pulled around with it as the planet rotates. In Jupiter's radiation belts, the trapped particles are ten times more energetic than those in the magnetic belts of Earth, and many times more abundant. Jupiter's magnetosphere contains an internal source of material in the volcanic activity of the moon Io [192]. Io is an important source of material for the magnetosphere and most of the particles in the magnetosphere come from Io. The energy of the magnetosphere comes from Jupiter's rotation. Ions and electrons are initially energized as they are spun up to co-rotation by the magnetic field into the Io torus. The magnetosphere is very dynamic. The magnetic equator is offset 10° to the equatorial equator. The magnetic field itself is not fully symmetrical (Fig. 5.3) but generally can be described as a dipole, tilted 9.6° to the rotation axis [194]. Therefore, being tilted the magnetosphere wobbles as it is whirled around by the planet's rotation. So, when viewed from a stationary point, it oscillates and twists. The outer fringe is affected by gusts from the solar wind and can change on a time scale of hours. There are also changes in the Io torus from year to year, perhaps due to changes in the moon's volcanic activity.

According to Rogers, the magnetic fields can be pictured as 'lines of force' like elastic cables running through space, anchored in the rotating planet. As the magnetic fields are stronger near the poles, the lines of force are closer together, the field lines converging towards the magnetic poles. Consequently, the field can be described as a magnetic bottle, and this is why it can accumulate plasma in the

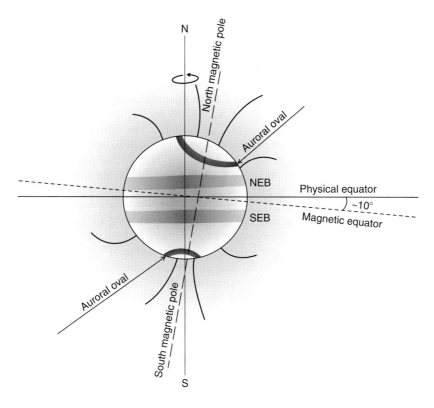

Fig. 5.3. A simplified view of Jupiter's offset, tilted magnetic poles, magnetic field lines, and auroral ovals. (After Rogers 1995 [193]).

form of high-energy 'radiation belts', both around Jupiter and around Earth [195]. We can imagine magnetic field lines by remembering how iron filings behave in the vicinity of a simple magnet. Remember the experiment in school? The teacher places a magnet on a table, and then pours iron filings around it. The iron filings organize themselves around the magnet in the shape of, and along the magnetic field lines of, the magnetic field running through the north and south poles of the magnet. The magnetic field lines of a planet behave and look much the same way.

The field itself probably originates deep within Jupiter. It is suspected that the field originates from 'dynamo' circulation deep inside Jupiter, within the outer shell of the hypothesized metallic hydrogen that is thought to make up the mantle of the planet. Like Jupiter, the magnetic field rotates, but with a rotation period of 9 h 55 min and 29.71 s. This rotation period is defined as System III, which is also the rotation period that coincides with radio bursts from the planet that are synchronized to particular orbital positions of the moon Io [196]. The axis of rotation of the magnetic field is tilted 9.6° from Jupiter's polar axis and it is offset $0.12\,R_J$ (R_J is one Jupiter radius) from Jupiter's center. System III is not to be confused with System I and System II rotation periods.

Of course, the existence and scale of Jupiter's magnetosphere was known from Earth based radio observations. But the exact intensities of the magnetic field and particle populations, and the detailed structure and true overall extent could not be accurately discerned until the spacecraft flybys of the 1970s and 1980s. Over time, various spacecraft examining different regions of the magnetosphere have provided a better understanding of the magnetosphere, showing how much the structure changes over time, and revealing the importance of the moon Io as a source of material (Fig. 5.4) [197]. By the end of the Pioneer/Voyager era, we knew the following (1) An electric current of more than a million amperes flows along the magnetic flux tube linking Jupiter and Io. (2) A doughnut-shaped torus containing sulfur and oxygen ions surrounds Jupiter at the orbit of Io. This torus emits ultraviolet light, has temperatures of up to 100,000 K, and is populated by more than 1,000 electrons cm^{-3}. A region of 'cold' (i.e., forced to rotate with the magnetic field) plasma exists between Io's orbit and the planet. It has larger than expected amounts of sulfur, sulfur dioxide, and oxygen, all probably derived from Io's volcanic eruptions. (3) The Sun-facing magnetopause (outer edge of the magnetosphere where the solar wind meets the magnetosphere) responds rapidly to changing solar wind pressure, varying from less than 50 Jupiter radii to more than 100 Jupiter radii from the planet's center. (4) A region of 'hot' (i.e., not forced to rotate with the magnetosphere) plasma exists in the outer magnetosphere. It consists primarily of hydrogen, oxygen, and sulfur ions. (5) Jupiter emits low frequency radio waves (wavelengths of one to several kilometers). The amount of radiation is strongly latitude-dependent. (6) There exists a complex interaction between the magnetosphere and the moon Ganymede. This results in deviations from a smooth magnetic field and charged particle distributions that extend up to 200,000 km from the satellite. (7) About 25 Jupiter radii behind the planet, the character of the magnetosphere changes from the 'closed' magnetic lines to an extended magnetotail without line closure. This occurs as a result of downstream interaction with the solar wind. (8) Jupiter's magnetotail extends to the orbit of Saturn - more than 700 million km 'downwind' of Jupiter [199].

For perhaps the first time, the simultaneous observations by Cassini and Galileo gave scientists a glimpse of the actual shape and curvature of the bow shock and magnetopause when the spacecraft were 100 Jovian radii apart, with Galileo sunward of Cassini and outbound from Jupiter on its 29th orbit [200]. The Cassini flyby of Jupiter offered an unprecedented opportunity, since Galileo was still in orbit about

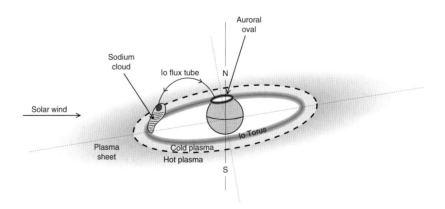

Fig. 5.4. A simplified view of the Io Torus, the Io Flux Tube, Sodium Cloud, Plasma sheet, and Polar Aurora. (After Rogers 1995 [198]).

the planet. As Cassini approached, one spacecraft characterized the solar wind, the other sensing the response of the Jovian magnetosphere. Later, as the relative positions of the two spacecraft changed, the roles reversed. For a short time both spacecraft were in the magnetosphere. The two spacecraft obtained data that showed the Jovian magnetosphere being compressed as a result of an inter-planetary shockwave, generated by the Sun, that impinged upon it [201]. As Cassini was inbound in situ (i.e, in the environment of Jupiter), it made measurements of the upstream solar wind conditions while Galileo was deep in the magnetosphere. Later, when Cassini was outbound moving past Jupiter, the roles reversed as Galileo was out in the solar wind and while Cassini crossed in and out of the magnetosheath [202].

Jupiter's interaction with the solar wind is not passive by any means. Its interaction is actually very active, and its magnetosphere leaves a distinct impression on the solar wind, even far upstream. The volcanoes of the moon Io supply heavy ions of mostly oxygenated sulfur to Jupiter's magnetosphere. According to Hill, once in the magnetosphere, these ions are energized and can escape magnetic confinement if they are neutralized through charge-exchange interactions. These energetic neutral atoms form a planetary wind that can actually move upstream against the solar wind. These ions are not coupled to the solar wind because they are not charged. Eventually, these energetic neutral atoms are re-ionized and do become part of the solar wind. However, because the solar wind is mainly composed of hydrogen like the Sun, the heavy ions retain the distinctive signature of their origins in the volcanoes of Io; thus, providing clear evidence of the Jovian magnetosphere's impact on the solar wind [203].

Remember that the magnetosphere is composed of the outer zone, the middle zone, and the inner zone (Fig. 5.2). The magnetic field in the outer magnetosphere is weak and variable on timescales of an hour or less, especially near the magnetopause. However, the plasma here is very rarefied and hot. Just inside the magnetopause it is the hottest thermal population known anywhere in the solar system (even the Sun), at 300–400 million Kelvin. However, the particles are very sparse. It is the pressure of this super hot plasma, and not the magnetosphere, that actually holds back the solar wind [204]. As the solar wind encounters the magnetosphere it is deflected around the planet. The magnetosphere is compressed on the sunward side, and as the solar wind slips around it, the magnetosphere is drawn out into a long tail. This magnetotail is quite long and can be 150–200 R_J. The "tail" can break off and travel "downwind" from the planet in the solar wind. Pioneer 10 and Voyager 2 encountered pieces that had broken off as far out as the vicinity of Saturn's orbit. The orbit of the Galileo spacecraft was maneuvered so that the spacecraft could explore the magnetotail, which it passed through at a distance of 143 R_J [205].

The middle zone extends from the Io plasma torus out to a distance of ~30–50 R_J. This part of the magnetosphere is dominated by the plasma sheet which is ~5 R_J thick [206]. According to Harland, the Galileo spacecraft demonstrated that the disk-like plasma sheet extends at least 100 R_J from Io's torus from which it derives, and that the thickness of the undulating disk varies significantly from one 10-h rotation to the next [207]. The plasma sheet is approximately equatorial and co-rotating although not exactly so. The plasma sheet is hot, although not nearly as hot as the hot plasma in the outer zone. Thus we refer to the plasma sheet as the 'cooler plasma.' Io and its neutral atomic cloud supply the Io plasma torus, which diffuses outward to form the plasma sheet. The composition of this denser but cooler plasma sheet is dominated by sulfur and oxygen from Io. In fact, most of the particles of the magnetosphere come from Io. The Io torus is the densest part of the magnetosphere and occupies a space starting at ~5.5 R_J and extending to 8 R_J [208]. By comparison, the orbital radius of Io is 5.9 R_J.

The inner zone extends from 5.5 R_J to 1.5 R_J. This region is dominated by Jupiter's intrinsic magnetic field. Here the total plasma density decreases but the highest energy particles reach their peak intensities [209]. In fact, inside the inner zone a human would receive a lethal dose of radiation in less than a minute.

To review, Jupiter's magnetosphere is basically composed of three zones. On the sunward side, the outer zone starts at the magnetopause and extends inward toward Jupiter to ~50 R_J. The middle zone extends from the outer zone inward to the Io torus at 8 R_J. The inner zone is dominated by the Io torus, which extends from 8 R_J to 5.5 R_J.

To further understand the structure of the magnetosphere, it can be helpful to examine the many parts of the magnetosphere as though we were in a spacecraft traveling to Jupiter from the direction of the Sun.

As we travel along with the solar wind, we find that we are suddenly compressed with other particles as we approach the vicinity of Jupiter. Here a bow shock develops. As we pass through the bow shock, we enter a relatively thin region called the magnetosheath. Soon, we cross an imaginary line called the magnetopause where the solar wind actually meets the magnetosphere and is deflected around and past the planet. We are now in the rarified plasma known as the outer zone. Next, we reach the equatorial, co-rotating plasma sheet of the middle zone. Finally, we reach the inner zone and the Io torus. Leaving this, we move back into the middle zone followed by the outer zone. These zones are somewhat stretched out on the side away from the Sun. And finally, we move into the magnetotail, which is the magnetosphere on the side away from the Sun that is drawn out by the solar wind. According to Rogers, the magnetotail starts ~150 R_J from the planet, where plasma breaks away from the co-rotating plasma sheet and flows down the tail [210]. Since the magnetotail has been detected all the way to the vicinity of Saturn, we see from our travel that Jupiter's magnetosphere truly is an enormous structure!

5.2 The Io Cloud and Torus

The core of the magnetosphere, its densest part, and the source of most of its plasma, is the Io torus [211]. Indeed, Io itself plays a significant role in Jupiter's magnetosphere, as it is believed that most of the particles in the magnetosphere come from Io. Lou Frank of the University of Iowa says the Io torus is the "beating heart of the Jovian Magnetosphere [212]." Material is transported from Io to the torus in neutral atomic form via a cloud of neutral atoms that is localized around Io itself [213].

This sodium cloud surrounding Io was actually discovered from Earth by Brown in 1972, being detected by the optical emission from the sodium atoms in it. Other atoms are also present in the cloud and these atoms come from Io [214]. It was previously known that the Io torus contained sulfur, oxygen, sodium, and potassium. In 1999, astronomers at Kitt Peak National Observatory in Arizona announced that the torus also contained chlorine [215]. The Io atomic cloud is localized around Io and is generally elongated in a shape resembling that of a banana (Fig. 5.4) [216]. While Io and the atomic cloud orbit Jupiter at 17 km s^{-1}, the plasma sheet co-rotates at 74 km s^{-1}, overtaking the cloud. Of course the atoms in the cloud are also in orbit around Jupiter. Those that diffuse inward toward Jupiter orbit faster and those that diffuse outward orbit slower. As the co-rotating plasma overtakes the cloud, impacts cause atoms to split into ions and electrons that can then be captured by the plasma sheet where they eventually migrate out into the magnetosphere [217].

If you imagine a giant tire inner tube encircling Jupiter's waist, or perhaps a really big doughnut, you get a pretty good picture of the Io plasma torus. The torus resides at 5–8 R_J, co-rotating with the magnetic field. Io passes through the torus only twice per orbit around Jupiter, since the torus is tilted 7° to the orbit of Io [218]. The torus is not a physically hard, impenetrable structure. It is not to be confused with Jupiter's gossamer rings. Rather it is a torus of ionized gas, i.e., a plasma, a halo that extends all the way around Io's orbit [219]. Pioneer 10 discovered the torus in 1973 when the spacecraft detected its ultraviolet emission. Later in 1979, Voyager determined the full extent of the torus [220]. The Pioneer 10, Voyager 1, and Ulysses spacecraft have all traveled through the torus. More recently the Galileo spacecraft has penetrated the torus. As a result, we know there are three concentric parts of the torus. There is a cold, inner torus at 5.0–5.6 R_J, inside Io's orbit. It receives only 2% of the plasma from Io [221]. There is the torus peak at 5.7 R_J, just inside the orbit of Io. This is the sharp boundary between the cold inner torus and the warm outer torus [222]. And there is the warm outer torus that resides at 5.9–7.5 R_J, at and outside Io's orbit. The outer torus is the main

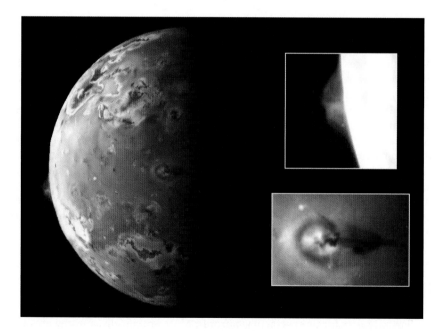

Fig. 5.5. Active volcanic plumes on Io. This Galileo spacecraft image shows two volcanic plumes on Io. One plume was captured on the bright limb of the moon over a caldera named *Pillan Patera*. This plume is 140 km (86 miles) high. The second plume is seen near the terminator (boundary between night and day) and is called *Prometheus*. The shadow of the 75 km (45 mile) high airborne plume can be seen extending to the right of the eruption vent. The vent is near the center of the bright and dark rings. North is toward the top of the image. (Credit: NASA/JPL-Caltech).

pool of plasma that gradually diffuses outward. As it does, it gains energy from the magnetic field and from the hot plasma that surrounds it [223].

There is a distinct outer boundary to the plasma torus where it comes up against the plasma sheet of the middle magnetosphere. From 7.5–9 R_J the plasma density falls off sharply and the temperature increases dramatically, fivefold [224].

We do not fully understand all the mechanisms that maintain the torus. It had been thought that neutral atoms were sputtering from the surface crust of Io as they were being bombarded by charged particles circulating in Jupiter's magnetosphere. But the more likely process became evident only when volcanic plumes were spotted on Io's limb (Fig. 5.5). The Voyager spacecraft saw ultraviolet emission from ionized oxygen and from both single and doubly ionized sulfur in the torus. Apparently, the plasma density is related to Io's volcanic activity [225]. By the time the Galileo spacecraft was into its mission, scientists had concluded that this is in fact the case. The Jovian magnetic field snatches away and then progressively accelerates away the ionized material ejected from Io [226].

Although the low escape velocity on Io allows the volcanic plumes to rise to several hundred kilometers in altitude, the eruptions do not occur with enough velocity to launch particles into Jupiter's magnetosphere. Instead, particles in the plumes become positively or negatively charged, making them susceptible to Jupiter's magnetic field, which grabs them into the magnetosphere. These particles form the doughnut shaped torus [227].

5.3 Radiation Belts

Like Earth, Jupiter possesses a system of radiation belts that function in much the same way. However, it was not known initially whether the radiation belt was a single belt or multiple belts.

The Galileo atmospheric probe carried an energetic particle instrument used to measure the fluxes of electrons, protons, helium nuclei, and heavy ions in the electromagnetic environment. Once the probe was inside Io's orbit, it started its energetic particles sampling. Scientists had expected the region inside the orbit of Jupiter's rings to be quiet. However, at a distance from Jupiter of 50,000 km, the probe detected a belt of intense radiation with a particle density ten times stronger than Earth's Van Allen Belts! The radiation intensities measured by the probe on its way in clearly indicated the presence of two distinct radiation belts, the inner of which had not been previously known [228].

5.4 Aurora

Like Earth, Jupiter forms spectacular auroras. More specifically, auroras are seen near and around Jupiter's poles, and occur in much the same fashion as they do on Earth. Aurora form an oval around each magnetic pole, where ions or electrons from the magnetosphere stream down into the ionosphere. These particles come from the Io orbital plasma cloud and beyond (Fig. 5.6). The aurora oval roughly marks the ring of the magnetic field lines that intersect Io's orbit and often has hotspots inside it [229].

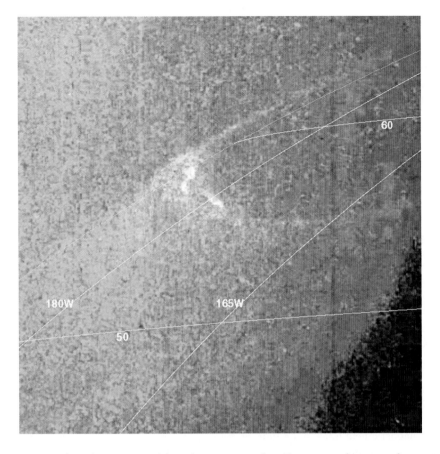

Fig. 5.6. False color aurora. A false color composite of Galileo spacecraft images of Jupiter's northern aurora on the night side of the planet. The glow is caused by electrically charged particles impinging on the atmosphere from above. The particles travel along Jupiter's magnetic field lines, which are nearly vertical at this latitude. The auroral arc marks the boundary between the "closed" field lines that are attached to the planet at both ends and the "open" field lines that extend out into planetary space. The colored background is light scattered from Jupiter's bright cresent. (Credit: NASA/JPL-Caltech).

The existence of Jovian auroral activity was first inferred from the particles and fields data collected by the Pioneer probes. The International Ultraviolet Explorer spacecraft, launched in 1978, made the first direct observations. The Galileo spacecraft later searched for auroral activity on the dark side of the planet, using its ultraviolet spectrometer [230]. Jupiter's auroras are best detected at far-ultraviolet wavelengths, where atoms and molecules of hydrogen radiate. These emissions can only be detected outside Earth's atmosphere. Consequently, the spacecraft missions of the 1970s and 1980s first detected the aurora in Jupiter's cloud tops, with the Earth orbiting observatory Copernicus being the first. Jupiter's aurora can also be detected from the Hubble Space Telescope (HST), as was first done in 1992 [231], and with modern infrared technology from ground based Earth instruments. Several research groups have made infrared images of Jupiter's aurora [232]. Earth orbiting spacecraft have also detected X-ray emissions coming from Jupiter's aurora regions [233]. The aurora can also be detected from Earth in the infrared at 3.5 μm. In far ultraviolet and infrared, the auroral oval can be detected in both the day and night side [234]. Auroral activity is concentrated in the oval ribbons, one around each magnetic pole [235]. It is thought that 'hazy clouds' are produced directly beneath auroral activity [236].

Jupiter's aurora is the most powerful in the solar system. The main features being the main oval, or footprint, that generally co-rotates with the planet, and a region of patchy, diffuse emission inside the oval on Jupiter's dusky side [237]. Jupiter's aurora is powered largely by energy extracted from planetary rotation, although there also seems to be a contribution from the solar wind. This contrasts with Earth's aurora, which is generated through the interaction of the solar wind with Earth's magnetosphere [238].

There is evidence that the auroral activity on Jupiter can flare and then subside rapidly. During a two hour period on September 21, 1999 the HST was used to make four imaging runs of Jupiter's northern far-ultraviolet aurora using HST's Imaging Spectrograph (STIS) in time-tagged mode. During a 4-min segment of the 2-h imaging run, a dramatic, rapidly intensifying, flare-like auroral emission was detected on Jupiter. The event began as a small 'pinpoint' emission near 167° System III longitude and 63° north latitude, which rose rapidly in intensity and became a structure several thousand kilometers across. The event apparently reached its maximum intensity in approximately 70 s, and then began to decrease in intensity and size. This entire sequence happened inside the four-minute segment. During this flare event, other auroral emissions remained virtually unchanged in both intensity and morphology. This flare event occurred within a region of fainter more diffuse emissions inside the main auroral oval [239].

The portion of the flare poleward of the main oval demonstrates that it is linked by magnetic field lines to a region of the magnetosphere lying at distances greater than 30 R_J on Jupiter's dayside. An analytical model indicates that the flare maps to a longitudinally extended region located at distances between 40 and 60 R_J in the morning sector. Therefore, it can be assumed that the flare was triggered by a disturbance in this region of the magnetosphere [240]. Calculations based upon measurements taken by the Advanced Composition Explorer spacecraft at the Earth L1 Langrarian position, indicates a series of sharp rises in the solar-wind dynamic pressure at the orbit of Jupiter near the time that the flare was observed [241].

We have already discussed the effect of the solar wind on the shape and size of the Jovian magnetosphere. We have also discussed the effect of the solar wind on changes in the intensity of auroral emissions. It is speculated that a sharp jump in the solar-wind dynamic pressure at Jupiter's dayside magnetopause produced the disturbance that manifested itself in the polar-cap flare. Solar-wind conditions at the time of the flare were not unusual, suggesting that such flares, if triggered by changes in the solar-wind pressure, may not be uncommon. The response of Earth's aurora to solar-wind dynamic pressure pulses are evidenced by rapid global brightenings associated with the passing of interplanetary shocks, and as smaller scale transient auroral events. Similar events may be at work at Jupiter [242].

During the Cassini spacecraft flyby of Jupiter, as Cassini was enroute to Saturn, scientists took advantage of the encounter to take data on Jupiter's auroral activity. In order to correlate changes in the intensity and morphology of Jupiter's aurora with the state of the solar wind, Cassini carried out a coordinated observation campaign with the HST and the Galileo spacecraft in orbit around the planet. Inbound, Cassini monitored the state of the solar wind, Galileo observed magnetosphere properties from within the magnetosphere, and HST observed Jupiter's aurora. Outbound, Cassini observed the night side aurora and traversed the edge of the magnetosheath, monitoring fluctuations in the shape and width of Jupiter's magnetosphere, while Galileo monitored the solar wind and HST monitored the dayside aurora [243]. The Cassini spacecraft observed three interplanetary shocks due to activity in the solar wind as it approached Jupiter from Earth. A distinct brightening in Jupiter's aurora followed each of these shocks. This proved that the interaction of the solar wind would in fact result in such brightenings [244, 245].

Jupiter's aurora has been observed to vary on short (minutes to hours) and long (days to weeks) timescales. This variability is thought to be due to the combined influences of internal magnetosphere processes and external solar wind driven changes. Similar to the processes driving the Earth's Aurora, a direct relationship between injection of electrons and a transient auroral feature was observed. Unlike the solar wind driven aurora at Earth, Jupiter's auroral morphology shows dependence on both the solar wind and Jupiter's rotation [246].

We know that energetic electrons and ions that are trapped in Earth's magnetosphere can suddenly be accelerated towards the planet. Some of the dynamic features of Earth's aurora (the northern and southern lights) are created by the fraction of these injected particles that travel along magnetic field lines and strike the upper atmosphere. The aurora of Jupiter resemble those of Earth in some respects; for example, both appear as large ovals encircling the poles and both show transient events. But the magnetosphere of Earth and Jupiter are so different in the way they are powered, that it was not known whether the magnetospheric drivers of Earth's aurora also caused them on Jupiter. Mauk et al., were able to demonstrate a direct relationship between Earth-like injections of electrons in Jupiter's

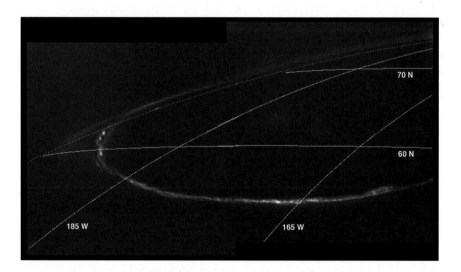

Fig. 5.7. Night side Jovian aurora. Galileo spacecraft image. The upper bright arc is auroral emission seen "edge on" above the planetary limb with the darkness of space as a background. The lower bright arc is seen against the dark clouds of Jupiter. The aurora is easier to see on the night side of Jupiter because it is fainter than the clouds when they are illuminated by sunlight. North is at top of image. (Credit: NASA/JPL-Caltech).

magnetosphere and a transient auroral feature in Jupiter's polar region. The discovery of Earth-like charged particle injections within Jupiter's magnetosphere is surprising because Jupiter's magnetosphere is powered mostly from the inside by the rapid but steady planetary rotation rather than from the outside by the variable solar wind. Mauk et al., took advantage of the Cassini flyby in late 2000 and early 2001, and the joint Cassini, Galileo, HST observation campaign and discovered the previously unknown role of injection in generating auroral emissions at Jupiter [247]. According to Scott Bolton, a member of the Galileo spacecraft plasma spectrometer team, the auroral arc on Jupiter is thin and patchy, contrary to those on Earth. Jupiter's auroral arc is estimated to occur between 300 and 600 km above the 1 bar reference level [248].

Three of Jupiter's moons, Io, Ganymede, and Europa, each produce a very distinct auroral "footprint" at Jupiter (Fig. 5.7) . The footprint of Io is by far the brightest, since its volcanic activity puts heavy ions into the Jovian magnetosphere [249, 250]. These footprints will be more fully discussed in the sections on The Io Flux Tube and Magnetic Footprints on Jupiter.

5.5 Radio Emission

Radio emissions from Jupiter provided the first clues that the planet had a strong magnetic field and a large magnetosphere [251]. Jupiter is a very strong radio source, especially at a wavelength of 30–0.6 MHz. Radiation in the radio part of the spectrum was first detected from Jupiter in the 1950s. Soon it was discovered that the radio bursts seemed to happen when certain longitudes were near Jupiter's central meridian. However, these radio bursts did not appear to be associated with any physical feature seen on Jupiter. Finally, in 1964 E. K. Biggs realized that they corresponded to particular orbital positions of the moon Io. Once again we see the strong influence of Io on Jupiter's electromagnetic environment [252].

These 30–0.6 MHz emissions are known as *decametric* emissions, and are only one of four classes of radio emission. We know that these decametric radio waves are "synchrotron radiation that spiral around the magnetic field lines" [253]. Synchrotron radio emission is produced when electrons are trapped in a magnetic field. The other three emissions are *decimetric, millimetric,* and *kilometric* emissions.

Decimetric emission is detected at wavelengths between 10 cm and several meters. These come from the inner magnetosphere at 1.3–3 R_J, and appear strongest at 1.5 R_J. This is the highest energy part of the radiation belts [254].

Millimetric emission is part of the thermal spectrum of Jupiter's atmosphere, detected at wavelengths up to 10 cm [255].

Kilometric emission can only be detected from space, and this emission from Jupiter was not discovered until Voyager reached the Jovian system [256].

There is evidence that Jupiter's radio emissions and aurora are controlled by the solar wind. Radio emission in wavelengths of 0.3–3 MHz is often referred to as *hectometric* radiation. Research by Gurnett et al., [257] indicates that the emissions are triggered by interplanetary shock propagating outward from the Sun. The same electrons that produce the hectometric emission also produce aurora. Thus, aurorae are also controlled by the interplanetary shock in the solar wind. It is believed that Jupiter's radio emissions are generated along high-latitude magnetic field lines by the same electrons that produce Jupiter's aurora, and both the radio emission in the hectometric frequency range and the aurora vary considerably. Evidence already existed that the solar wind plays a role in controlling the intensity of hectometric radiation. The Cassini and Galileo spacecraft confirmed this effect on Jupiter's hectometric radiation and aurora. Simultaneous, coordinated observations of Jupiter's hectometric radio emission and extreme ultraviolet auroral emissions were carried out by the Cassini and Galileo spacecraft on 30 December, 2000, during the Cassini fly-by of Jupiter. This observation revealed that both of these emissions were triggered by interplanetary shocks propagating outward from the Sun [258]. We also learned that Jupiter's synchrotron emission can vary on relatively short time scales, even as short as a matter of days [259].

When an interplanetary shock interacts with Earth's magnetosphere, we know that the magnetosphere becomes strongly compressed. This results in a large-scale reconfiguration of the magnetic field and associated acceleration and energization of plasma in the magnetosphere. The stresses associated with the compression lead to large field-aligned currents and electric fields, particularly along the high-latitude magnetic field lines where the electrons that carry the field-aligned currents strike the atmosphere and produce the aurora. It is inferred that similar processes are responsible for the Jovian hectometric radiation and aurora [260]. The observations made by the Cassini and Galileo spacecraft present strong evidence that Jovian hectometric radiation and Jovian auroral extreme ultraviolet emissions are triggered by interplanetary shocks [261]. Thus, even radio emission from Jupiter is affected by interplanetary shock waves in the solar wind.

5.6 Lightning

Spacecraft missions in the 1970s and 1980s revealed that Jupiter experiences lightning in its clouds much as Earth does. W. Borucki and M. Williams (1986) calculated that the lightning must occur at the 5-bar level, far below the clouds visible to us, in giant thunderstorms with the same dynamics as those on Earth. Indeed, spacecraft detected lightning photographically on Jupiter's night side, and heard the lightning as 'whistlers' with a plasma wave experiment [262]. More recently, the Galileo atmospheric probe detected lightning when it descended into the cloud tops of Jupiter. This probe 'heard the radio bursts of 50,000 lightning flashes during the 57.6 minutes of its descent' [263].

I think it is fascinating that lightning occurs on a planet other than Earth. The fact that the physical conditions that cause lightning can exist on another planet so much unlike Earth is truly a wonder. But, there is an important reason for studying the lightning on Jupiter. Lightning is diagnostic of dynamics, chemical composition, and heat exchange within the Jovian atmosphere [264]. Although lightning on Earth is evenly distributed geographically, lightning on Jupiter appears to be confined to high latitudes [265]. On Earth we experience ground strike lightning but also discharges between clouds, which are far more common. Since Jupiter has no 'ground', lightning there is presumed to be from cloud to cloud. The average lightning on Jupiter is ten times stronger than the average lightning on Earth [266].

The Galileo spacecraft searched for lightning on Jupiter's dark side and detected a chain of flashing thunderheads just south of the westward moving jetstream at 46° north latitude (Fig. 5.8). Almost all the lightning detected by Voyager had also been detected near the latitude of a westward moving jetstream. The lightning was

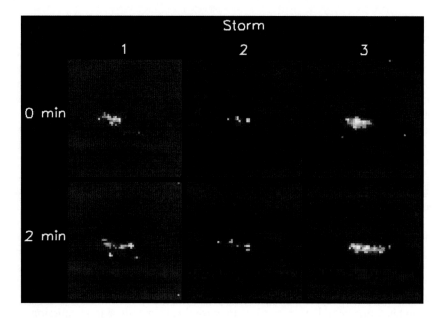

Fig. 5.8. Changing Lightning Storms on Jupiter. Galileo spacecraft images showing lightning storms in three different locations on Jupiter's night side. Each panel shows multiple lightning strikes, coming from different parts of the storm. The lightning originates in Jupiter's water cloud, which is 50–75 km (30–45 miles) below the ammonia cloud. The latter acts as a translucent screen, diffusing the light over an area proportionate to the depth of the lightning. The lightning strikes are hundreds of times brighter than lightning on Earth. The bottom row shows the same three storms as the top row, but the bottom-row images were taken two minutes later. The panels are 8,000 km on a side. North is at the top of the picture. (Credit: NASA/JPL-Caltech).

flashing far below the visible ammonia cloud deck that was acting like a translucent screen, diffusing the light upward. The apparent width of the flash could be used as an indicator to infer its depth of 75 km, a depth that would be consistent with water clouds [267]. Previously, the Voyager observations determined that Jupiter's lightning occurred at a depth of ~5 bar in water clouds [268].

Voyager observed that the Great Red Spot (GRS) spawned a series of rapidly expanding white clouds resembling large thunderheads, which were carried away by the prevailing jetstream. Indeed, a telescope of just 6-in. aperture can reveal the large bright wake on the following side of the GRS. Later, the Galileo spacecraft detected lightning in these GRS-following 'thunderheads' and confirmed they are intensely convective features. Galileo made multi-filter observations of a cloud in one of these amorphous cyclonic regions, and the observation of lightning was consistent with this observation (Fig. 5.9). Multiple lightning strikes confirmed that this was a site of moist convection in the saturated environment at a depth of 75 km. The bright cloud Galileo observed resembled a site of convective upswelling in Earth's atmosphere [269]. As Cassini passed Jupiter on its way out to Saturn, it also made observations of Jovian lightning. Its observations of lightning in the Hα line of the lightning spectrum also suggested that the lightning Cassini observed was at depths of more than 5 bars in the atmosphere [270]. We know there is great turbulence following the GRS. The most interesting lightning seen by the Galileo

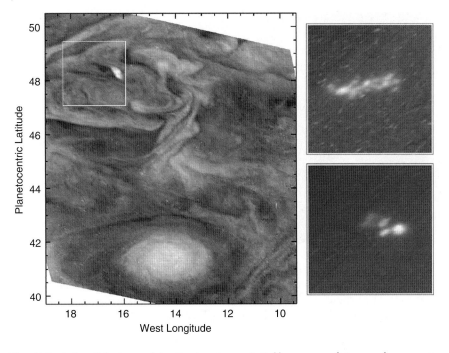

Fig. 5.9. Jovian Lightning and the Daytime Storm. A Galileo spacecraft image of a convective storm (*left panel*) and the associated lightning (*right panels*) in Jupiter's atmosphere. The *left image* shows the daytime view. The *right images* show the area highlighted (box) in the dayside view as it appeared 110 min later during the night. Multiple lightning strikes are visible in the nightside images, which were taken 3 min and 38 s apart. The bright, cloudy area in the dayside view is similar in appearance to a region of upwelling in Earth's atmosphere. The dark, clear region to the west (*left*) appears similar to a region of downwelling in Earth's atmosphere. (Credit: NASA/JPL-Caltech).

spacecraft occurred in the GRS wake. Apparently, this region of unusually strong turbulent eddies in the atmospheric flow is favorable for unusually strong lightning. One of the four storms observed by Cassini resided in the turbulent wake of the GRS [271].

Jupiter's lightning appears to occur in clusters, also called storms (Fig. 5.10). The Galileo spacecraft observed multiple flashes in these storms and that the storms are separated by large distances. Most of the storms observed by Galileo occur in cyclonic shear zones. The only exceptions are the storms between 40° and 50° north, which seem to be clustered near the center of westward jets [272]. Both the Voyager and Galileo spacecraft observed considerably fewer storms in the southern hemisphere of Jupiter.

Since lightning on Jupiter is expected to occur in water clouds, lighting deeper than 5 bars suggests that the water clouds themselves exist deeper than 5 bars on Jupiter. Thus, we see demonstrated the benefit of observing Jupiter's lightning, as no other investigative technique can observe much below the 5 bar level on Jupiter. To support water clouds at depths more than 5 bars and the corresponding temperature associated with that depth, water abundance in the deep Jovian atmosphere should be more than 1× solar abundances [273, 274].

The Cassini spacecraft observed four lightning spots on Jupiter as it passed by, and these were correlated with four unusually bright, small (~1,000 km in size)

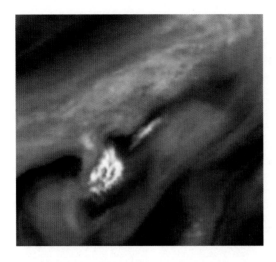

Fig. 5.10. Water Cloud Thunderstorm on Jupiter. A false color image of a convective thunderstorm northwest of the GRS taken by the Galileo spacecraft. The white cloud in the center is a tall, thick cloud 1,000 km (620 miles) across, standing 25 km (15 miles) higher than the surrounding clouds. (Credit: NASA/California Institute of Technology/Cornell University).

clouds that Cassini had previously observed a few hours earlier on the day side. Apparently the bright, small clouds are quite rare, with only a few on the planet at any given time. Visible and near infrared (IR) spectra suggested that these clouds were dense, vertically extended, and contained unusually large particles. This is also typical of terrestrial (Earth bound) thunderstorms [275]. Lightning in the Voyager 2 observations were not always correlated with small bright clouds. Because Cassini at the distance it passed could observe very few and only the most powerful lightning storms, the correlation of the clouds with Cassini lightning may suggest that only the most powerful thunderstorms with bright lightning will penetrate the troposphere up to the levels where they are easily observable in reflected light [276]. The Cassini observations also confirmed the Voyager conclusions that Jovian thunderstorms may generate lightning even when the clouds do not extend to the top of the troposphere and expose themselves as bright clouds. Such deep storms may develop the high tops several hours or even days after or before the deep thunderstorm stage, or they may remain at deep levels through their whole life span [277].

We can now be fairly certain that Jovian lightning occurs from cloud to cloud discharge, as most lightning occurs on Earth, and that Jovian lightning occurs in water clouds. The Jovian lightning gives us an indication of the depth of water clouds on Jupiter and, by estimating the temperature at these depths allows us to make an estimate of the Jovian water abundance ratio compared to the solar ratio. We have learned much from the spacecraft missions mentioned above; yet, we have so much to learn about Jovian lightning and the mechanisms that drive it.

5.7 The Io Flux Tube and Magnetic Footprints on Jupiter

As we have discussed, Jupiter and its moon, Io, are strongly linked in an electromagnetic fashion. One of the examples of this connection is a phenomenon called "the Io flux-tube." Io is electrically linked to Jupiter by a pair of "flux-tubes" that run down the magnetic field lines in Jupiter's polar regions [278].

The Io flux tube is a fascinating feature of Jupiter's electromagnetic environment and John Rogers explains it very well. According to Rogers, "As the plasma-filled magnetic field sweeps past Io, it induces an electrical potential across Io which drives an electric current, either through its body, or more likely, through its ionosphere. Plasma sweeping past Io tends to be entrained to the orbital speed of Io and accelerated north or south into this current, which becomes a so-called 'flux tube' of particles running along the magnetic field lines that intersect Io. The flux tube is actually guided by a wave that propagates from Io along the field lines, called an 'Alfven wave'. On the side of the flux tube toward Jupiter, electrons stream away from Io toward the ionosphere, and ions stream in the opposite direction. On the side away from Jupiter, the directions are reversed" (Fig. 5.4) [279].

Io also leaves a magnetic footprint on Jupiter's upper atmosphere. This footprint appears as a spot of ultraviolet emission that remains fixed underneath Io's position as Jupiter rotates. We know that the magnetic footprint of Io extends much further than the immediate vicinity of the Io flux tube interaction with Jupiter [280]. There is also faint, persistent, far-ultraviolet emission from the footprints of Ganymede and Europa. The emissions of the magnetic footprints of Io, Ganymede, and Europa were detected in ultraviolet images taken with the HST Space Telescope Imaging Spectrograph (STIS). The fact that these footprints are associated with the respective Jovian moons was established by observing that the footprint remains stationary under its moon as Jupiter rotates. Ganymede's footprint appears distinctly brighter that Europa's, and Io's is the brightest of all [281, 282].

The emission of each footprint persists for several hours downstream. That is, the ultraviolet emission signature gives the appearance of a comet, with the footprint stationary under its moon being the head, and the fading downstream emission being the tail [283]. Harland describes the auroral footprints as 'having comet-like tails' because the charged particles continue to excite Jupiter's atmosphere for some time after Io has passed overhead. Of course, we understand that it is Jupiter that rotates so rapidly under the moon as the moon slowly orbits (Fig. 5.7) [284].

There is a tremendous amount of energy in the Io flux-tubes. On December 7, 1995, the Galileo spacecraft flew past Io on its way into orbit around Jupiter. As it passed Io, the spacecraft detected bi-directional electron flows aligned with the Jovian magnetic field lines. This was the first in situ observation of the Io flux tubes. The flow of particles was equivalent to an electric current of several millions amps and an overall deposit of energy into Jupiter's atmosphere of a trillion watts; the most powerful direct current in the solar system! That is incredible energy of which the human mind can hardly comprehend! This energy generated the auroral emissions that appear as footprints in the ultraviolet, as it strikes the Jovian atmosphere [285]. "Io's downstream emission extends for at least 100° in longitude along the magnetic footprint of Io's orbit". Clarke et al., believe this implies active processes that persist for a few hours after Jupiter's magnetic field has swept past Io. Clarke et al., conclude that these downstream emissions are produced by high-energy charged particles that precipitate into Jupiter's atmosphere from the plasma torus downstream from Io, a process that continues at a declining rate for several hours after Io has passed [286]. These footprints are evident at both Jovian poles, as is the auroral polar hoods, discussed previously. The exact cause of these emissions is not fully understood.

5.8 X-Ray Emission from Jupiter and Its Environ

Next to our Sun, Jupiter is the strongest and most interesting X-ray source within our solar system. Observations with the Chandra X-ray Observatory (CXO) and the XMM-Newton Observatory revealed that Jupiter's environment is a rich source of X-rays and that the structure is very complicated. There appear to be four distinct sources of X-ray emission (1) the high-latitude auroral zones, or polar auroral zones on Jupiter; (2) the disk of Jupiter; (3) the Io plasma torus; (4) and the Galilean moons [287].

Elsner et al., explain the production of X-rays as: "A number of interactions among electrons, protons, ions, neutral atoms, and electromagnetic fields lead to the production of X-rays. One of the simplest is the interaction between an electron and proton or ion leading to the emission of a photon. The electron begins and ends unbound to the heavy positively charged particle. This process is called *bremsstrahlung* and leads to broadband continuum emission. For sufficiently energetic electrons, the spectrum can peak in the X-ray band or at higher frequencies" [288].

X-ray emission in Jupiter's auroral regions is attributed to charge-exchange between energetic ions and neutral atoms high in the polar atmosphere. X-ray emission from Jupiter's moons may result from the energetic ions incident on their surfaces ionizing and exciting neutral surface atoms leading to fluorescent K-shell line emission [289]. High spatial resolution observations of < 1 arcsec of Jupiter with the CXO in 2000 found that most of the auroral X-rays were located in small high-latitude regions, and in the north were confined to longitudes between 160° and 180° System 3, and in latitude between 60° and 70°. This correlated strongly with the Jovian magnetic field and mapped along magnetic field lines to distances greater than 30 Jupiter radii (R_J) from the planet. CXO observations in 2003 of the southern hemisphere found auroral emissions to be more extended in longitude than in the north, but still tied to magnetic field lines. Astronomers conclude that the X-ray auroral emission regions reside well within the ultraviolet (UV) auroral oval (discussed earlier). Typically the auroral regions emit ~0.5–1.0 GW in soft X-rays [290].

Gladstone et al., call this concentration of auroral X-rays a "pulsating auroral X-ray hot spot." Comparison of Chandra X-ray emission mapping with simultaneous (December 18, 2000) far-ultraviolet images obtained by the HST imaging spectrograph revealed that the northern auroral X-rays are concentrated in a "hot spot" within the main ultraviolet auroral oval at high magnetic latitudes. Gladstone et al., agree with Elsner et al., that the hot spot is located at ~60–70° north latitude and ~160–190° system III longitude. No southern latitude hot spot has been detected, but that may be due simply to the poor viewing geometry of the southern polar cap during these observations [291].

Astronomers have also concluded that the emission from the auroral region is variable. Short, seconds to minutes, UV flares have been observed to be accompanied by an X-ray flare, with the location of the X-ray flare slightly offset from the location of the UV flare. An unexpected observation of a ~40-min periodic oscillation in the X-ray emission from Jupiter's northern auroral zone was detected by the CXO in December 2000. Surprisingly, the February 2003 CXO observations did not find a 40-min oscillation. However, CXO did detect variability on a timescale from 10 to 100 min. In February 2002, the Ulysses spacecraft observed 40-minute oscillations that were correlated with periodic radio bursts from Jupiter. Although at the time of the 2003 CXO observations, the Ulysses radio observations did not detect periodic 40-min oscillations, but did detect variations on timescales similar to that in the X-ray emission [292]. Astronomers do not yet fully understand the relationship of X-ray emissions to other forms of emission.

Astronomers have also established beyond a doubt that the bright X-ray emission from the Jovian polar-regions is line emission, not a continuum, and is likely caused by charge exchange between energetic highly-stripped heavy ions and neutral atoms in Jupiter's upper atmosphere [293]. According to Elsner et al., the CXO X-ray spectrum for the northern region of Jupiter indicates strong evidence that one of the major constituents of the incoming ion stream is highly ionized oxygen. There must be at least one other major constituent to account for the line emission. The two strongest candidates are highly ionized sulfur and highly ionized carbon. Sulfur would favor a magnetospheric origin while carbon would favor a solar origin. The CXO data seem to favor sulfur. However, XMM-Newton spacecraft data from April 2003 seems to favor carbon. Astronomers have not been able to arrive at a conclusion [294].

The planetary disk of Jupiter seems to be awash in X-ray emission. CXO data indicates Jupiter's X-ray emission appears to be a featureless disk illuminated by solar X-rays [295], or from a combination of reflected and fluoresced solar X-rays [296]. This seems to be a rather simple picture, but astronomers continue to investigate other possible causes of X-ray emission from Jupiter's disk.

The Io Plasma Torus is also an emitter of X-rays. According to Elsner et al., CXO observations in November 1999 and December 2000 detected a faint diffuse source of soft X-rays from the region of the Io plasma torus. Apparently, knowledge of the spatial structure in X-rays and X-ray spectra of the Io plasma torus are limited by its faintness. So far, Na, Cl, S, and O ions and possibly protons, have been detected in the Io plasma torus X-ray spectrum [297].

X-ray emissions have also been observed from the Galilean moons, and how these emissions occur is of great interest to astronomers. In 1999 and 2000, the CXO detected X-ray emission from the Galilean moons, specifically X-ray emissions from Io and Europa. The estimated power of the emission from Io was 2 MW, and 3 MW for Europa. X-ray emission from Callisto is suspected but was not detected by CXO, apparently occurring below the spacecraft's sensitivity level during the observation. According to Elsner et al., "the X-ray emissions from Europa are best explained by energetic H, S, and O ion bombardment of the icy surface with subsequent fluorescent emission from the deposition of energetic particle energy in the top 10 μm of the surface" [298].

Thus, the environment of Jupiter is a rich, diverse source of X-ray emission. While our knowledge of Jovian X-ray emission is limited, perhaps future spacecraft missions in orbit around Jupiter making in situ observations will eventually yield the data astronomers long for, as advocated by Elsner et al. [299].

5.9 Summary

As we now understand, there is so much more to Jupiter and the Jovian system than can be seen in visible wavelengths. The electromagnetic environment of Jupiter is one of the largest, most complicated structures in the solar system.

And although the study of Jupiter's electromagnetic environment is presently the domain of professional astronomers, an understanding of this environment gives us a more complete picture of what is going on with Jupiter. And who knows, can the day be far off when amateurs will be making these observations, or at least mining the data collected by others? The availability of technology is advancing fast, so stay tuned.

Chapter 6

The Jovian Satellite System

Jupiter is practically its own solar system, only lacking its own heat source, like the Sun. And yet, the tidal forces it exerts does result in some heat related phenomena in some of its larger moons. In addition to the four large Galilean moons, Jupiter also has a family of smaller regular moons and irregular moons. In fact, as of November 27, 2006 there were 63 known moons of Jupiter (Scott Sheppard, personal communication). Jupiter has a ring system, undetected until the spacecraft missions of the 1970s. There is also a family of asteroids known as Trojans, and a Jupiter family of comets. As we will see, Jupiter has substantial gravitational influence in our solar system.

6.1 The Galilean Moons of Jupiter

Galileo first discovered Jupiter's four largest moons when he observed Jupiter with a telescope on January 7, 1610. These four largest moons are named Ganymede, Europa, Callisto, and Io. Of course, now we know these 'Galilean' moons are interesting little worlds in their own right. Perhaps 'little' is an inappropriate description, since Ganymede, Callisto, and Io are each larger in diameter than Earth's Moon.

Albedo markings were detected on all the Galilean moons with large telescopes and good seeing long before the spacecraft missions of the 1970s, although great detail was never detected and the observations were always suspect. The Voyager spacecraft missions detected a wealth of detail on the surfaces of these moons, along with other kinds of data. The Galileo spacecraft mission that followed provided even greater detail and a new cache of data that scientists will be studying for years. Yet, there are many interesting phenomena that amateurs can observe with modest telescopes. Since the Galilean moons orbit Jupiter in its equatorial plane, ground based telescopes can observe transits of the moons in front of the planet. And, we can observe the shadows of these moons on Jupiter's surface, appearing ink black against Jupiter's bright cloud tops. We can observe eclipses of the moons as they disappear and reappear in and out of Jupiter's shadow. We can also observe occultations of the moons by Jupiter, as they disappear behind the limb of the planet and reappear on the other side. About every six years, when the Jovian system is edge-on to Earth, we can see the moons occult and eclipse each other. Mutual occultations of the moons offer one of the only chances to detect color differences between the moons. Schedules for all of these events can be found at various scientific web sites and in schedules published monthly in Sky and Telescope and Astronomy magazines. And, with the great advances in CCD cameras and web cams, now even experienced amateurs are capturing images of Jupiter's moons that show amazingly detailed albedo markings. Yes, it can be very rewarding to observe Jupiter and its moons and to show them to a person for the first time. The movement of the moons in relation to Jupiter can be detected in a relatively short period.

The orbits of the Galilean moons are eccentric, or non-circular. The Galilean moons deform the others orbits, while Jupiter's tidal forces try to circularize them. Io is most affected by this tug of war since it is closest in to Jupiter. The effect of all of this is to cause tidal heating of Io and Europa [300].

6.1.1 Io

Io is a fascinating world! We have already seen in Chap. 5 that Io is a major polluter of the Jovian environment. Mainly due to volcanic activity, Io creates clouds of neutral atoms, sheds dust, deposits sulfur on the surfaces of its neighboring moons, and supplies charged particles to the Io torus. Indeed, the Galileo spacecraft detected significant variability in the concentrations of dust and charged particles from one spacecraft orbit to another, probably due to varying intensities of volcanic activity [301]. What a busy moon! As we shall see, Io displays some of the most intriguing volcanic related activity in the solar system [302]! Astronomers were greatly surprised when the first close look at Io revealed that its surface was devoid of craters (Fig. 6.1)! We now know that this is due to the great geologic activity on this moon and the constant resurfacing of its surface. Much new data was collected during the Galileo era and scientists have delved into it to further our understanding of Io. Galileo was launched in October 1989 and ended by plummeting into Jupiter in September 2003.

Io's diameter is 3,642.6 km [303] and it is so close to Jupiter that it completes an orbit around its parent in just 1.769 days [304]! It has a mean density of

Fig. 6.1. High Resolution Global View of Io by the Galileo Spacecraft. Io is the most volcanic body in the solar system. Its surface reveals rugged mountains, layer materials forming plateaus, and many irregular depressions called volcanic calderas. Several of the dark, flow-like features correspond to hot spots, and may be active lava flows. There are no impact craters seen, as volcanism covers the surface with new deposits more rapidly than impacts from comets and asteroids can occur. (Credit: NASA/JPL/University of Arizona/PIRL)

3.5294 ± 0.0013 g cm^{-3} [305]. The tidal force of Jupiter acting on Io melts its interior, drives its volcanoes, covers its surface with sulfur, provides all the wonderful colors on its surface, and spits out a transient atmosphere of sulfur and sodium that makes its way into the Io torus cloud [306]. Yes, due to the tidal forces caused by its close proximity to Jupiter, Io is a tortured world. Io's surface is pockmarked with calderas, lava lakes, and plumes. Its volcanic activities are not evidenced so much by pyroclastic clouds, like volcanoes on Earth, but rather geysers similar to "Old faithful" in that they are concentrated and powerful. These 'geysers' are referred to officially as 'plumes' on Io, and they are not ordinary by any sense of the word (Fig. 6.2).

However, volcanic eruptions are not the only event on Io. During a 5-year period, the Galileo spacecraft recorded 82 surface changes on Io, from various processes [307]. Io is so close to Jupiter that differential gravity has distorted the moon's shape. The surface of the hemisphere that faces Jupiter forms a 100-m tall bulge [308]. Io's atmosphere is tenuous at best. Being composed of SO_2, its atmosphere can reach 10-7 bar at midday or under volcanic plumes, but is completely frozen out at night. Surface temperature on Io can range from 120–130 K at midday to 60–90 K at midnight [309].

The plumes on Io are fascinating. During the Galileo spacecraft mission, two types of plumes were observed. These two types of plumes are categorized based upon their size and the colors of their deposits, or fallout. According to Geissler et al.,

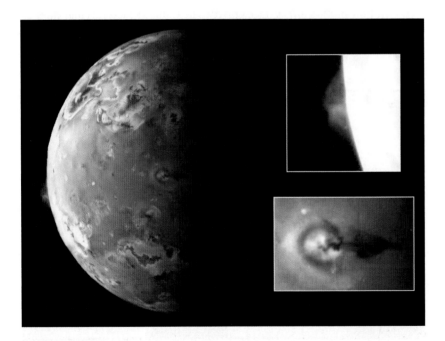

Fig. 6.2. Active volcanic plumes on Io. The Galileo spacecraft capture two active plumes on Io. One plume is clearly seen on the bright limb, or edge, of the moon. This plume is 140 km (86 miles) high. *Prometheus*, the second plume is, is seen near the terminator. The shadow of the 75-km high plume can be seen extending to the right of the eruption vent. The vent is near the center of the bright and dark rings. (Credit: NASA/JPL/Caltech)

smaller plumes produce circular rings averaging 150–200 km in radius that are yellow or white in color. Evidence indicates that these plumes are the most numerous ones and normally coat their surroundings with frosts of fine-grained SO_2. The repeated eruptions of these smaller plumes are most likely significant contributors to Io's resurfacing rate. Larger plumes are much less numerous, but produce orange or red sulfur rich rings, oval in shape, that typically range in diameter from 500 to 550 km (Fig. 6.3). The larger plumes probably account for most of the dust ejection. Both types of plumes can occur as a single episode, or can be somewhat continuous over a period of time [310]. It is theorized that the two classes of plumes are driven by different volatiles. The smaller plumes are powered by the explosive volatilization of SO_2. These plumes contain silicate ash and or sulfur compounds that cause the discoloration of the resulting deposits. The larger plumes probably contain significant quantities of sulfur, evidenced by extensive red or orange deposits [311]. According to Harland, Susan Kieffer's comprehensive analysis and model showed that sulfur dioxide at a depth of several kilometers would be heated to 1,400 K upon direct contact with silicate lava. Once boiled under pressure, the gaseous sulfur dioxide would force open pre-existing cracks in the crust and vent explosively upon reaching the surface [312].

Galileo observed four volcanic eruptions that were described as explosive [313]. Some of the larger plumes can be truly spectacular. A plume seen by the Galileo spacecraft over the volcano *Tvashtar* in late 2000, rose ~400 km high [314] (Fig. 6.4)! Galileo later detected an even larger plume that rose ~500 km over a feature provisionally named *Thor*. Apparently these plumes rise so high because of the intensity of their eruption and Io's low surface gravity. The Galileo spacecraft may have actually flown through the plume of *Thor*, since the Galileo Plasma Subsystem detected SO_2 near closest approach at this time [315]. Due to the great height of these plumes,

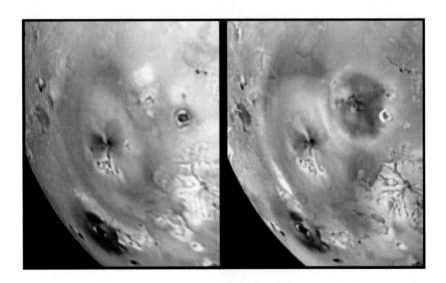

Fig. 6.3. Arizona-sized Io eruption. The right frame was captured by the Galileo spacecraft five months after it captured the left one. During this time a new dark spot 400 km in diameter, roughly the size of Arizona, developed in the right half of the frame around a volcanic center named *Pillan Patera*. (Credit: NASA/JPL/University of Arizon/PIRL)

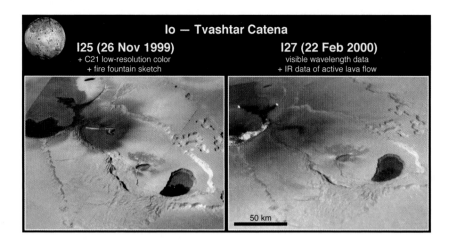

Fig. 6.4. Eruption at Tvashtar Catena on Io imaged by the Galileo spacecraft. This pair of images captures a dynamic eruption at Tvashtar Catena, a chain of volcanic bowls on Io. A change has occurred in the location of hot lava over a period of a few months in 1999 and early 2000. This style of volcanism on Io is unpredictable and short-lived. (Credit: NASA/JPL/University of Arizona/PIRL)

a large surface area can be affected. Late in its mission, Galileo detected a large plume deposit ring south of the *Karei* area. The interior radius of this ring was 415 km, and its exterior radius averaged 690 km. The area covered by the deposit was almost 933,000 km² [316]!

As noted, plumes can occur on a continuous basis. Both Voyagers 1 and 2 observed prominent plumes coming from a feature known as *Masubi*. During the Galileo mission, distinct surface changes were noted since the Voyager observations, and Galileo observed two distinct eruptions that produced prominent rings around *Masubi*. The first eruption occurred between orbits 9 and 10, leaving a prominent dark ring. However, by the next observation on orbit 11, the rings had begun to fade. Eight months later, on orbit 15, the rings had completely disappeared. So, the rings from the fallout can be quite short-lived [317].

The most prolific plume on Io appears to be from the feature known as *Prometheus*. This most persistently active plume was observed by both Voyagers 1 and 2, and by the Galileo spacecraft on every favorable encounter. According to Geissler et al., *Prometheus* produces a dusty plume 50–150 km high that deposits a bright ring of SO_2 ~250 km in diameter, as well as fainter rings both interior and exterior to the bright ring. The center of the bright ring had migrated 85 km to the west since the Voyager encounters. The plume apparently derives from the distal end of lava flows issuing from a fissure, and these flows had grown in length since Voyager. The plume either moved along with the flow, or another plume developed in a different place along the flow (Fig. 6.5). We see that not only can activity occur continuously, but it can also move around contributing further to the surface changes on Io [318].

Another continuous plume is one associated with *Pele*, having been first observed by Voyager 1. Due to its consistent high-temperature thermal emission, giant plume, and enormous red ring, *Pele* is one of Io's most distinctive plumes (Fig. 6.6). The Hubble Space Telescope detected plumes here in 1998, 1996, and

Fig. 6.5. Sources of Volcanic Plumes Near *Prometheus*. *Prometheus* is the "Old Faithful" of Io's many active volcanoes. A plume has been seen here everytime the viewing conditions were favorable for the Voyager and Galileo spacecraft. However, the location of the vent that is the source of the plume is still a mystery. The lava flow extends 90 km from the source. Bright patches probably composed of sulfur dioxide can be seen in several places along the flow's margins. Galileo spacecraft image. (Credit: NASA/JPL/University of Arizona/PIRL)

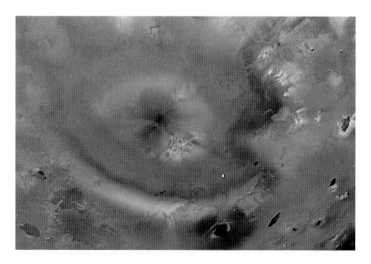

Fig. 6.6. A Voyager 1 image taken in March 1979 looking straight down at *Pele*, one of Io's most active plumes. The heart-shaped feature is where the material in the plume falls to the surface. (Credit: NASA/JPL/Caltech)

1997. Even the Cassini spacecraft detected a plume over *Pele* as it passed Jupiter on its way to Saturn [319]. Of the total resurfacing caused by plume activity *Pele*, because of its size and activity, accounts for over 40% of it. While some areas of Io, such as that around *Pele*, were resurfaced continually, much of Io remained unchanged during the course of the Galileo mission. Approximately 83% of Io's visible surface never changed during the Galileo era [320].

While evidence suggests that plumes normally eject SO_2 and sulfur, evidence has also been found of silicate deposits. In fact, taking images with red, green, and violet filters, Galileo detected deposits of SO_2, sulfur, and silicate [321]. In July 1997, the Galileo spacecraft found a 200 km tall plume rising from *Pillan Patera*. Later, a gray colored plume deposit was seen around *Pillan*, indicating the ejected material had been mostly silicate, not just sulfur [322]. Some plume deposits are irregular in shape, while others such as those from *Prometheus*, produce well-defined rings [323].

Although the giant plumes do not contribute much to the resurfacing of Io, Scientists believe that they contribute significantly to the escape of dust from Io into the Jovian environment because of the high velocity at which the dust is ejected. And, even though the plumes are tenuous, they easily supply enough mass to account for the flux of dust escaping Io. These plumes provide a connection between the geologic activity on the surface of the moon and the flux of materials into space. Plumes eject dust and gas directly and help to sustain the tenuous atmosphere that is eroded by impacting charged particles [324]. The repeated eruptions of smaller SO_2 dominated plumes do contribute significantly to Io's resurfacing rate [325]. The majority of eruptions deposit materials that can reach distances between 50 and 350 km. A much smaller class of giant eruptions can deposit materials as far away as 350–800 km.

Ultimately, Galileo data allowed scientists to conclude that there are two distinct types of plumes. Smaller plumes produce near-circular rings typically 150–200 km in radius, white or yellow in color, that can be contaminated with silicates, and that frequently coat their surroundings with frosts of fine-grained SO_2. Galileo observed a much smaller number of larger plumes, which produced oval orange or red, sulfur rich rings with maximum radii in the north–south direction that typically range in radius from 500–550 km [326].

While plumes may be the most exciting events on Io, surface changes are also caused by other events. Numerous small-scale changes have been documented using Voyager and Galileo spacecraft data. Many paterae were seen to brighten or darken during the Galileo mission. One assumption is that these changes were caused by thermal activity, heating the surfaces of the paterae, causing the existing SO_2, a bright substance, to sublimate away. Another assumption is that the thermal activity heated the SO_2 surrounding the paterae, causing it to liquify the ice, allowing it to flow onto the paterae floor flooding and brightening it. Or that fresh lava flowed from the interior due to thermal heating, darkening the surface of the paterae with new material [327] (Fig. 6.7). High resolution imaging of *Amirani* showed that new lavas covered an area ~620 km^2 during a 134-day interval. The lavas apparently erupted from 23 separate locations across flow fields. Lava flows were also detected at *Prometheus* and numerous other locations. Scientists believe the Galileo data revealed the overturning of lava lakes, evidenced by the eruption of fresh lava onto surfaces already covered by lava. Thus, we see three main classes of surface changes: volcanic plume deposits, patera color or albedo changes, and SO_2 seepage [328].

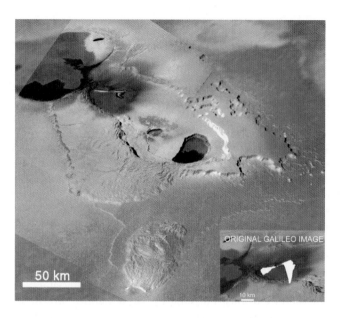

Fig. 6.7. The Galileo spacecraft caught Tvashtar Catena in active eruption on the surface of Io in November 1999. The molten lava was so hot that it saturated, or over-exposed, Galileo's camera. The lava appears to be producing fountains up to 1.5 km above the surface! Several other lava flows can be seen on the floors of the calderas, with the darkest flows probably the most recent. (Credit: NASA/JPL/University of Arizona/PIRL)

Even from a single site, eruption styles can be diverse. During the Galileo mission, *Ishtar Catena* exhibited many eruption styles, including a curtain of lava fountains, extensive surface flows, and a ~400 km high plume. These events occurred over a relatively short period of time, ~13 months [329].

After the Voyager missions, it was thought that most of the volcanic flows were due to sulfur flows. However, with the Galileo mission scientists began to realize that the temperatures of flows were so high that they had to be silicate based. Many of the flows observed were too hot to be sulfur based, since sulfur would vaporize at ~700 K. Galileo measured a lava flow at *Prometheus* at 1,100 K that had to be silicate rich [330]. On the other hand, *Emakong Patera* is surrounded by lava flows that are notable for being yellow-white. It is thought that this lava flow had been moderately hot sulfurous lava rather than very hot silicate. So, there is till evidence for sulfur flows, but silicate now seems to be more prominent [331]. One important result of the Galileo mission is that Io's paterae appear to be persistent lakes of lava [332].

During the Galileo era episodic brightenings that have been interpreted to be SO_2 seepage occurred at several locations around Io; specifically at *Haemus Montes, Zal Montes, Dorian Montez,* and the plateau to the north of *Pillan Patera.* These are regions with pronounced topography adjacent to active volcanic centers [333]. A brightening of a bright halo around *Haemus Mons* apparently occurred between Voyager 2 and Galileo's first orbit around Jupiter. This brightening was attributed to possible SO_2 seepage from the base of the massif [334].

There were also many smaller-scale changes in color or albedo that were confined to patera surfaces; such as those located at *Gish Bar, Itzamna, Camaxtli, Kaminari, Reiden, Pillan, Dazhbog,* and *Amaterasu.* Most of these locations are recognized hot spots, and in some cases the paterae later produced major eruptions that changed the surrounding surfaces for hundreds of kilometers [335]! Often, Io's volcanoes give notice they are about to erupt by a darkening of their caldera. Such events are common on Io, along with patera brightenings and color changes. This may signify transient heating within a caldera [336].

As with Earth, lava flows on Io can be associated with extremely high temperatures, and there may be a correlation between extremely high temperature and plume activity. There have been four well documented eruptions with temperatures exceeding 1,400 K on Io: *Pillan, Masubi, Pele,* and *Surt.* We can probably assume there have been many other events simply not caught by Galileo or other observational means. Scientists believe that such high temperatures may indicate the presence of lava fountaining, which is driven by volatiles that also produce large plumes in Io's tenuous atmosphere [337]. Galileo observed a spectacular eruption at Tvashtar, an eruption occurring along a ~25 km long fissure that produced a fire-fountain of lava reaching 1 km in height! This eruption was also observed from Earth with the IRTF and Keck AO systems. These ground based observations helped establish the ~36 h duration and 1,300–1,900 K temperature of the fire-fountain event [338].

The most surprising thing learned during the Galileo mission may be that there were only a small number of volcanic centers that visibly altered their surroundings. Out of over 100 active volcanoes and 450 paterae considered young and potentially active, only 28 produced noticeable surface changes more than a few tens of kilometers in extent. Few of the high temperature thermal events detected by Galileo were accompanied by plume events, and even the powerful eruptions of *Loki* left no visible mark on the surrounding terrain [339] (Fig. 6.8).

Mass wasting processes act on Io; slumping and landslides occur in close proximity to each other, thus there is spatial variations in material properties over distances of several kilometers (Fig. 6.9). However, even though there is a lot of evidence for mass wasting events, the floors of paterae lying close to mountains are relatively free of debris. Thus, the rate of volcanic resurfacing seems to dominate [340]. Apparently, Io is so volcanically active that it is resurfaced at a rate of ~1 cm per year [341, 342]. In fact, Io is the most active known planetary body in terms of luminosity and resurfacing rate, even more active than the Earth [343]!

Io's volcanic activity does not tend to build mountains or domes and in fact there are few mountains or mountain ranges on Io. We do not see the mountain building associated with plate tectonics as we see on Earth. But Io does have mountains. One of them is *Haemus Mons*, a ~10 km high mountain surrounded by a bright halo of SO_2 [344]. At its base, *Haemus Mons* appears to span 200 km [345].

Io's mountains are isolated, stand alone, and do not appear to form mountains ranges (Fig. 6.10). Thus, they do not appear to be the result of large-scale techtonism as on Earth. Io's mountains are interesting because they rise directly off the plains without foothills. Their origin is a mystery, since sulfur does not have the strength to form peaks [346]. Galileo images show its mountains to be collapsing under their own weight. They must therefore be a relatively recent creation [347].

Paul Schenk and Mark Bulmer, using stereo images taken by Voyager in 1979, postulate that mountains like *Euboea Mons* are individually faulted and uplifted blocks of crustal material [348] (Fig. 6.11). The mountains are massifs that could

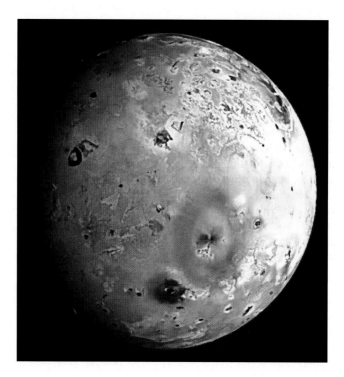

Fig. 6.8. A Galileo spacecraft image showing active volcanic centers on Io. *Loki Patera* is the large, dark horsehoe shaped feature close to the terminator toward the south pole of the moon. The big, reddish-orange ring in the lower right is formed by material deposited from the eruption of *Pele,* Io's largest volcanic plume. North is at the *top* of the picture. (Credit: NASA/JPL/University of Arizona/PIRL)

Fig. 6.9. Slumping Cliff on Io. The Galileo spacecraft caught this image of a mountain named *Telegonus Mensa* on Io showing outward slumping. The cliff is slumping due to gravity. The sun illuminates the surface from the upper right. (Credit: NASA/JPL/University of Arizona/PIRL)

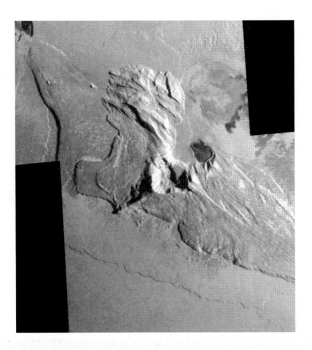

Fig. 6.10. A Galileo spacecraft image reveals details around a peak named *Tohil Mons,* which rises 5.4 km above Io's surface. Few of Io's mountains actually appear to be volcanoes. However, the shape of the pit directly to the east of *Tohil's* peak suggests a volcanic origin. (Credit: NASA/JPL/ University of Arizona/PIRL)

have been upthrusted recently – thus they are not necessarily ancient. The underlying structure of the mountains appears to be rigid silicate blocks. These blocks are probably thrust upward and then tilt. The angular mountains appear to be younger than the rounded ones. There may be structural relationships between mountains and calderas on Io. The absolute ages of Io's mountains are unknown [349].

High-resolution images taken by Galileo also reveal unexplained ridges on otherwise flat terrain. Some sites of ridge formation can be attributed to down-slope motion of loose material. But, this cannot explain the ridges seen on plains that are relatively flat. These ridges are similar to dunes on Earth and Mars. Earth and Mars have atmospheres with enough density to allow particle transport and thus dune formation. But the atmospheric pressure on Io is too low to do this. Io's winds cannot transport particles to form dunes [350].

These ridge, or dune-like, features are quite common on Io, since 28% of the Galileo high-resolution images show them. The regions with abundant ridges are rich in volatiles, dominated by SO_2. Thus, it is speculated that the presence of a blanket of volatiles is necessary for ridge formation [351]. Therefore, since Io lacks an atmosphere with enough density, scientists conclude that the formation of these ridges is consistent with formation from tidal flexing of Io by Jupiter, in the presence of a substantial volatile rich deposit on the surface [352].

Previous modeling supported the hypothesis that Io has a massive metallic core in which a magnetic field may be produced [353]. With the Galileo flyby of Io, the moment of inertia could be determined, allowing the model of Io's interior to be

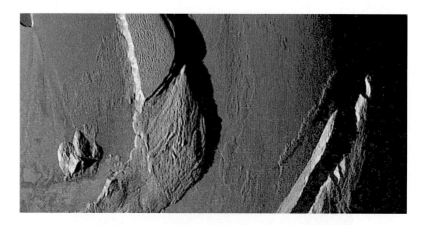

Fig. 6.11. Mountains of Io. The Galileo spacecraft caught this image of mountains of Io. The mountain just *left of center* in the image is 4 km (13,000 ft) high, and the small peak to the left is 1.5 km (5,000 ft) high. These mountains seem to be in the process of collapsing. (Credit: NASA/JPL/ University of Arizona/PIRL)

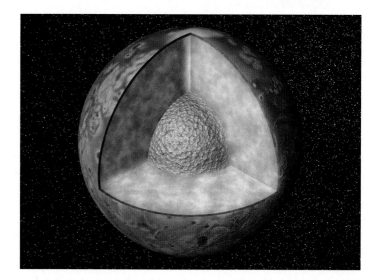

Fig. 6.12. This is a cutaway view of the possible internal structure of Io. The moon has a metallic iron, nickel core. The core is surrounded by a rock or silicate shell, which extends all the way to the surface. (Credit: NASA/JPL/Caltech)

refined. The low value that resulted from the flyby indicates that Io possesses a two-layer structure, with a metallic core most likely of iron and iron-sulfide, enclosed by a partially molten silicate mantle, over which there is a volcanically active crust, or lithosphere (Fig. 6.12). The Galileo data suggested that the core would be 36–52% of the moon's radius [354]. Io is so hot that its silicate lithosphere is very thin, and the magma is never very far from its surface [355].

6.1.2 Europa

Europa is the second closest Galilean moon to Jupiter, with an orbital period of 3.551 days [356]. It is the smallest of the Galilean moons with a diameter of 3,130 km and a mean density of 2.989 ± 0.046 [357]. Europa is intriguing for a number of reasons, not the least of which is that it may have a liquid ocean beneath its surface that scientists think might harbor some form of life!

The surface temperature on Europa is 120–131 K [358] and its atmosphere is tenuous, at best. The materials present on the surface are the products of radiation transformations, cryovolcanism, impact and gardening events that have occurred over time. Chemical alteration of the ice has been shown to produce condensed hydrogen peroxide within some icy surface regions. An atmosphere composed mostly of atomic and molecular hydrogen with some atomic and molecular oxygen from radiation processing of surface ice has also been detected [359]. As early as 1994, the Hubble Space Telescope had detected oxygen emissions from Europa [360]. The Cassini spacecraft Ultraviolet Imaging Spectrograph (UVIS) showed the presence of an extended oxygen atmosphere in addition to a bound molecular oxygen atmosphere. The UVIS observations also indicated the presence of atomic hydrogen and possibly other elements. Europa's water ice surface undergoes charged particle bombardment. This erodes the surface due to 'sputtering' and to a lesser extent sublimation. 'Sputtering' produces hydrogen and oxygen. The hydrogen is lost to space leaving the oxygen behind, thus the thin, bound oxygen predominant atmosphere and escaping hydrogen [361]. A torus of energetic neutral atoms has also been associated with Europa [362].

The surface of Europa has been characterized as 'billiard ball' smooth (Fig. 6.13). There is so little topographical relief that the vertical range of its surface is confined to a few hundred meters. Europa is covered with surface ice and though relatively

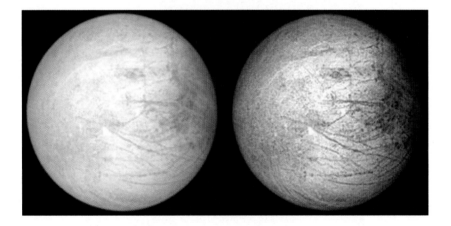

Fig. 6.13. This color composite view shows a view of the moon Europa in natural color (*left*) and in enhanced color (*right*). The bright white and bluish part of Europa's surface is composed mostly of water ice. The brownish, mottled regions on the right side of the moon may be covered by hydrated salts and an unknown red component. The yellowish mottled terrain on the left side is caused by some other unknown component. North is to the *top* of the image. (Credit: NASA/JPL/University of Arizona/PIRL)

smooth, this surface presents albedo, color, and texture variations [363]. This surface ice is thought by some to be a mixture of salt brines [364].

At low resolution, the surface can be classified into two main terrain types, ridged plains and mottled terrain. High-resolution Galileo images reveal that the mottled terrain consists predominantly of chaos regions. There is debate regarding the mechanisms that form these features, but the most popular model is that convection in the underlying ice shell deforms the surface, producing chaos regions, pits, and domes [365].

Figueredo and Greeley actually identify five principal terrain types: plains, bands, ridges, chaos, and crater materials. These are thought to result from tectonic fracturing and lineament building, cryovolcanic reworking of surface units with possible emplacement of sub-surface materials, and impact cratering [366] (Fig. 6.14). Miyamoto, *et al.*, describe additional surface features as domes, platforms, irregular uplifts, and disrupted micro-chaos regions. Some of these features show positive elevations reaching 100–200 m or more, and have surface textures bearing no relation to the surrounding terrain, These features appear to have obscured and spread over the pre-existing surface as a viscous flow; that is, as though there has been a fluid emplacement of ice or slush on the surface. These features are rarely disrupted by tectonic structures such as ridges and must therefore be some of the youngest features on Europa's surface [367].

Four types of plain are seen. 'Undifferentiated plains' are smooth, gradational with adjacent terrain, and are cut by numerous linear features. 'Bright plains' are located towards high latitudes, and are crisscrossed by a variety of linear features. 'Dark

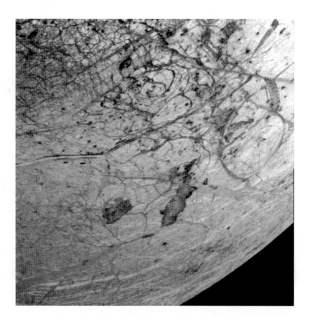

Fig. 6.14. The Galileo spacecraft caught this southern hemisphere image of Europa in February 1997, showing the southern extent of the "wedges" region, an area of extensive disruption. South of the wedges, the eastern extent of the Agenor Linea is also visible. Thera and Thrace Macula are the dark irregular features southeast of Agenor Linea. North is to the *top*. (Credit: NASA/JPL/Caltech)

plains' resemble the light plains but are darker. 'Fractured plains' give the appearance of being shattered and bear curved gray streaks and numerous brown spots [368].

The various linear features are what stand out when first looking at Europa. The 'triple bands' comprise a bright stripe, possibly a ridge, running down a dark band (Fig. 6.15). They run for thousands of kilometers but rarely exceed 15 km wide. They often start or end near dark circular spots, or brown mottled terrain. These may be caused by stresses induced by the moon's orbital eccentricity. The 'dark wedges' that are seen can be 300 km long and may be 25 km wide at the open end. Since they cut across older features, they appear to be the result of fractures, splitting and spreading the surface ice, with the gaps being subsequently refilled with slush that froze again [369] (Fig. 6.16).

Ridges are by far the most common linear feature on Europa. A double ridge feature consists of a central trough bounded by a ridge pair. In some instances, double ridges taper and continue as single ridges, troughs, or fractures. Single ridges are usually smaller than there double counterparts and form relatively short segments [370]. There are examples of linear features and ridges that show evidence of filling in or flooding, specifically swamping of preexisting ridges and grooves by a fluid that has erupted onto the surface. This is the first evidence of ice flows on any of the Galilean moons [371]. Another exciting discovery were features that scientists referred to as 'icebergs.' Galileo revealed that some of Europa's surface had been fractured into polygonal 'rafts' of ice that were individually 3–6 km across (Fig. 6.17). The chaos around these features was stained brown, suggesting the presence of minerals of endogenic origin. Scientist Ron Greeley said the blocks of ice were "similar to those seen on Earth's polar seas

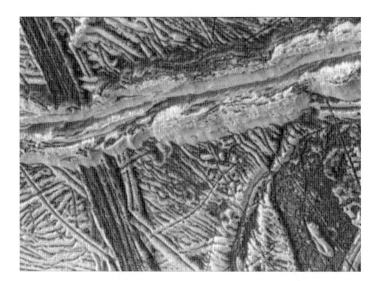

Fig. 6.15. This high-resolution image of Europa taken by the Galileo spacecraft shows a dark, relatively smooth region at the lower right corner that may be a place where warm ice has welled up from below. The image also shows two prominent ridges that have different characteristics; the youngest ridge runs from left to top right and is about 5 km wide. The ridge has two bright, raised rims and a central valley. It overlies, and therefore must be younger than, a second ridge running from top to bottom on the left side of the image. (Credit: NASA/JPL/Caltech)

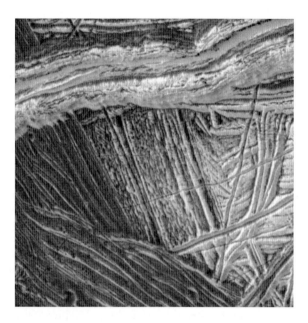

Fig. 6.16. Europa wedge region. A Galileo spacecraft image showing crustal separation, with the dark bands being areas where the icy crust has completely pulled apart. Dark material has welled up from below and filled the void created by this separation. (Credit: NASA/JPL/Caltech)

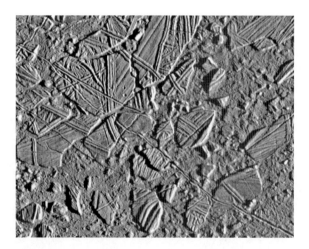

Fig. 6.17. This image, taken by the Galileo spacecraft, reveals the ice-rich crust of Europa. Crustal plates up to 13 km (8 miles) across have broken apart and "rafted" into new positions. (Credit: NASA/JPL/Arizona State University)

during springtime thaws." According to Paul Geissler the motion of the ice rafts could not be explained by convection in ice; only a fluid medium could account for such rotation and tilting. According to Michael Carr, the rafts had been clearly caught up in a strong current in a fluid medium, which was almost certainly water. Because they had been floating, they were true icebergs and most of their bulk would have been below the water level, just like on Earth (Fig. 6.18). Thus, we see strong evidence for some kind of liquid ocean. Carr believes the icebergs provide the proof that Europa had to have had liquid water exposed at the surface at some time in its past [372] (Fig. 6.19).

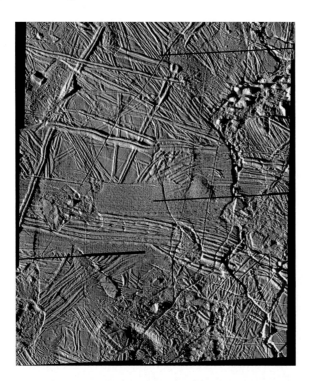

Fig. 6.18. This mosaic of images taken by the Galileo spacecraft reveals ridges, plains, and mountains on Europa. There are hundreds of ridges that cut across each other, indicating multiple episodes of ridge formation either by volcanic or tectonic activity within the ice. There are also numerous isolated mountains or "massifs". Irregular shaped areas where the ice appears to be lower than the surrounding plains may be related to the "chaos" areas of iceberg-like features. (Credit: NASA/JPL/Arizona State University)

Fig. 6.19. A very high resolution Galileo spacecraft image of the Conamara Chaos region on Europa. Icy plates have been broken apart and moved around laterally. There are corrugated plateaus ending in icy cliffs over 100 m high. (Credit: NASA/JPL/Caltech)

Bands are generally linear, curvilinear, or cycloidal features, with sharp parallel to subparallel margins. The interior of smooth bands consists of either very subdued ridges and troughs or material with little or no structure. It is common that the material forming the bands has a relatively lower albedo than the surrounding plains. Smooth bands are interpreted as regions of crustal extension in which the low relief and lack of internal structure can be due to small-scale fracturing or the emplacement of infilling material [373] (Fig. 6.20).

'Cycloids' are chains of scalloped lines linked arc-to-arc at their cusps, extending over the icy plains for hundreds of kilometers. Evidently, the cuspate form is the result of pressure on the underside of the thin ice shell induced by diurnal tides in the enclosed ocean. This caused Europa's ice shell to flex. Jupiter's immense gravity caused the ice to bulge, cyclically, over the period of an orbit. Once the tidal force exceeded the tensile strength of the ice, it began to crack. The crack propagated relatively slowly across the ever-changing stress field, following a curved path. As soon as the stress subsided, the fracture halted. Later, when the stress built up again, the crack restarted, but it did so along a new curve, and so on. This tidal cycle occurs about every 85 h. The distinctive scalloped appearance resulted from the fact that successive curved fractures shared cusps. The creation of these cycloids could

Fig. 6.20. This mosaic of Galileo images shows chaos and gray bands crossing Europa. The icy crust has been broken apart revealing a darker underlying material. The smooth gray band is an area where the crust has been fractured, separated, and filled in with material from the interior. Obviously, the moon has been subjected to intense geological deformation. (Credit: NASA/JPL/Caltech)

only happen if the tidal bulge could slide freely over the interior, implying there is a deep ocean separating the thin icy shell from the underlying silicate lithosphere. There is apparently no evidence that these cycloids are still being formed. The absence of new formation does not mean there is no longer a liquid ocean. It may simply mean that the icy crust has finally become thick enough that the stresses cannot overcome its tensile strength to fracture and create the cycloids. However, some scientists believe we should still be seeing current activity if a liquid ocean still exists. Obviously, if a new cycloid had been observed in the process of formation, it would be evidence that the ocean tides were still active [374]. According to Lee et al., cycloidal cracks are probably the most important among Europa's surface features for proving a subsurface liquid ocean, since the tidal stress that produces them can only approach the ice failure strength if an ocean is present [375].

'Chaos' features can be described as materials that form irregular areas containing blocks or polygons of preexisting crustal material, with other intervening material that appears to lie at the same or a lower level than the surrounding plains [376] (Fig. 6.21). Chaos regions seem to be most simply explained by melt-through events. A natural explanation is that the Chaos regions represent areas where deep, warm ocean waters have come into contact with the overlying ice. Partial melt-through events, which should be more common, may also allow disruption of the surface ice and formation of Chaos [377]. According to Greenberg et al., 18% of the surface of Europa is fresh appearing chaos while another 4% appears as slightly modified chaos. There is much more older chaotic terrain that is overprinted by tectonic structures. This chaos suggests that the effects of having liquid water under a very thin ice shell have dominated Europan geology. Chaos regions are widespread on Europa, and widespread chaos regions suggest that there has been occasional zero shell thickness [378]. Like craters, chaos areas may have been formed continuously over at least the entire geological age of the surface, leaving only the most recent

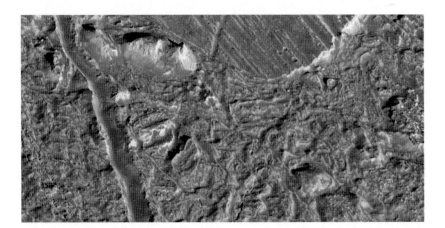

Fig. 6.21. This Galileo spacecraft view of chaotic terrain on Europa shows an area where the icy surface has been broken into many separate plates that have moved laterally and rotated. The plates are surrounded by a topographically lower matrix that may have been emplaced by water, slush, or warm flowing ice, which rose up from below the surface. One of the plates is seen as a flat, lineated area in the upper portion of the image. (Credit: NASA/JPL/Caltech)

chaos in recognizable form. Chaos may also have formed concurrently with ridge formation, as a long-term process [379].

Cryovolcanism is the eruption of liquid or vapor phases of water or other volatiles that would be frozen solid at the normal temperatures of an icy satellite's surface. Cryovolcanism has apparently resurfaced broad areas of Europa by disruption, displacement, and reworking of crustal ice [380] (Fig. 6.22).

There are craters on Europa's surface, but very few of them. The rarity of large impact craters suggests a geologically young surface, perhaps due to widespread resurfacing processes already alluded to [381]. The craters seem to fall into two categories. One type, referred to as 'palimpsests' are 100 km across with concentric fractures and low surface relief, as though the impact was later filled in. This 'filling in' leaves a very shallow crater or one that is almost not there. The second class consists of about a dozen craters with diameters of ~25 km (Fig. 6.23). The ice was probably thin at the time of impact. One crater in particular, *Pwyll Crater*, displays fantastically bright ejecta rays. These bright rays are fresh fine water-ice particles that were ejected and painted streaks for thousands of kilometers.

The bright rays indicate a very young crater. The lack of craters overall suggests that Europa's surface is relatively young [382]. According to Moore et al., Europa's craters appear to be anomalously shallow compared to similarly sized craters on other solid-surface bodies, which may be due to post-impact isostatic adjustment [383]. Thus, if there is a subsurface liquid water layer on Europa, the overlying ice must be thick enough that ~3–6 km deep craters do not penetrate completely [384]. Ruiz believes that analysis of size and depth of the largest impact structures suggests that these features were formed in an icy shell at least ~19–25 km thick [385]. Even the largest impact feature, *Tyre*, transported material to the surface from

Fig. 6.22. A wide-field image of a portion of Europa's icy surface, revealing ridges, plateaus, and patches of smooth, low-lying darker materials. Note the absence of craters, indicating this region is composed of a young surface material, suggesting that cryovolcanism has resurfaced the region. (Credit: NASA/JPL/Arizona State University)

Fig. 6.23. This enhanced color image from the Galileo spacecraft is of a young impact crater named *Pwyll*, a feature that is about 26 km in diameter. *Pwyll* is thought to be one of the youngest features on Europa. The bright white ejecta rays extend in all directions for over a thousand kilometers from the impact site. (Credit: NASA/JPL/University of Arizona/PIRL)

a depth not greater than ~4 km deep. And, *Pwyll* and *Manannan* transported material from a depth not greater than ~2 km [386]. Many crater floors lie at the same general elevation as the surrounding terrain beyond the continuous ejecta blanket, as opposed to being lower than surrounding terrain due to the excavation of the impact. This is true for craters *Cilix* and *Manannan* and may be a result of fluid fill in or isostatic adjustment following the impact [387]. Europa's larger craters, *Cilix*, *Maeve*, and *Pwyll* appear to exhibit central peaks, or central peak complexes. *Cilix* displays an elongated central peak complex surrounded by a flat crater floor, terrace walls, a circular rim, and reddish brown continuous ejecta blanket. The central peak complex is composed of two prominent massifs. Both massifs exhibit ~300 m relief, and is located in the center of the crater floor. The crater floor reveals only a few tens of meters relief and is mottled by a number of sub-kilometer reddish-brown patches [388]. Unlike the Moon or Mars, the central peaks of Europa's largest craters rise well above the surrounding crater rim. In the case of *Pwyll*, the central peak rises ~800 m above the crater floor, or ~300 m above the average rim height. By contrast, central peaks rising above rims are rare for the Moon and Mars [389].

Shoemaker estimated Europa's crater retention age as 30 Myr for craters less than 10 km, and found that the cratering record is consistent with a 10-km-thick ice crust overlying a liquid ocean [390]. There are ~150 impact craters less than 1 km across and these craters show no evidence of degradation by tectonic activity, again suggesting a decline in geologic processes on Europa, and that impact cratering was one of the last geologic processes to have occurred [391] (Fig. 6.24).

Impact into ice does not produce exactly the same effect as an impact into a silicate solid. Consequently, most craters on Europa are dissimilar to craters seen on Earth's Moon. Likewise, impact into solid ice produces an effect different from an impact into a low-viscosity material. Impact simulations suggest that the Europan features *Callinish* and *Tyre* would not be produced by impact into a solid ice target, but might be explained by impact into an ice layer of order of 10 km thick overlying a low-viscosity material [392]. It is believed that *Tyre* is a relatively recent cratering event [393].

It is important to understand the order in which resurfacing events have occurred on Europa. It is believed that tectonic resurfacing dominates the early formation of background plains by the intricate superposition of lineaments, the opening of wide bands with infilling of inter-plate gaps, and the buildup of ridges and ridge complexes along prominent fractures in the ice. Due to the lack of craters being overprinted by lineaments, it is thought that tectonic resurfacing decreased rapidly after ridged plains formation (Fig. 6.25). Later, the degree of cryogenic resurfacing increased with time. The transition from tectonic to cryogenic dominated resurfacing may be attributed to the gradual thickening of Europa's cryosphere. The evolution from a brittle ice shell to a thicker one would cause a decrease in fracturing or the melt through of sub-surface material due to tidal and endogenic processes; fracturing and plate displacements decreasing with time as the shell thickened [394].

Fig. 6.24. This Galileo spacecraft image shows the Conamara Chaos region, revealing craters that range in size from 30 m to over 450 m. The large number of craters seen here is unusual for Europa. This section of Conamara Chaos lies inside a bright ray of material that was ejected from the large impact crater *Pwyll*. The presence of craters within the bright ray suggests that many are secondaries which formed from the impact of chunks of material that were thrown out by the enormous energy of the impact which formed *Pwyll*. (Credit: NASA/JPL/Caltech)

Fig. 6.25. The Galileo spacecraft imaged this ancient impact basin on Europa. The "bulls-eye" pattern appears to be a 140-km wide impact scar that formed as the surface fractured minutes after a mountain-sized asteroid or comet slammed into the moon. This color composite represents a series of events. The earliest event was the impact that formed the Tyre structure. The impact was followed by the formation of the reddish lines superposed on Trye, the red color indicating areas that are probably a dirty water ice mixture. The fine blue-green lines are ridges that formed after the crater from west to east. (Credit: NASA/JPL/University of Arizona/PIRL)

The resurfacing of Europa did not occur as discreet events, but rather more likely as a continuous process, or stages. The first stage involved the sequence of formation and deformation of the background plains, a complex process. Tectonic processes were predominant at this stage, far outweighing cryovolcanism or impact cratering. The early second stage was probably entirely dominated by tectonic resurfacing. Tectonic processes also dominated the late second stage, when several sets of lineaments with very consistent orientation developed. During the third stage, resurfacing was primarily cryovolcanic and involved the formation of chaos and subdued pitted plains. The fourth resurfacing stage on the trailing hemisphere of Europa involved the development of the last sets of lineaments and the cryovolcanic formation of some chaos at middle southern latitudes. Lineaments include ridges and ridge complexes. On the trailing hemisphere the final events include the opening of linear and slightly cuspate regional fractures, and the formation of impact craters and associated ejecta deposits. On the leading hemisphere, the final stage involves the development of cuspate smooth bands and double ridges at high latitudes [395].

Even though the resurfacing process changed from tectonic-dominated processes to cryovolcanic-dominated processes, the transition appears to have been gradual, with both processes coexisting at most times but in varying degrees [396]. Kargel et al., believe that all of these features, their relationships, and their ages indicate that the crust of Europa has been completely resurfaced in the recent geological history [397].

Previous modeling supports the hypothesis that Europa has a massive metallic core in which a magnetic field may be generated [398]. Europa's core radii is estimated to be 426–510 km for an iron (Fe) core and 610–706 km if composed of iron and iron sulfide (Fe – FeS) [399] (Fig. 6.26). Thus, core size is between 10 and 45% of Europa's radius and contains up to 15% of the moon's mass [400].

Based upon Galileo data, modeling suggests that Europa is differentiated with a metallic core rich in iron (Fe), underlying a dehydrated silicate rock mantle and an ice – liquid shell ~120–140 km thick. The thickness of the water ice shell is uncertain and may range from less than 100–200 km. This water ice shell contains ~10% of the moon's total mass. Europa's weak magnetic field may be explained by either the existence of a metallic core or a liquid water ocean (beneath a thin solid shell) containing some electrolyte, or a combination of both [401]. There is indirect geological and geophysical evidence that Europa may posses a subsurface salty liquid water ocean [402]. Voyager and Galileo imaging and infrared spectroscopy show that Europa's surface is covered with water ice [403]. There is geological evidence that the water ice shell is decoupled from Europa's deep interior due to the existence of a subsurface liquid water ocean or at least a soft ice layer. The most convincing argument for a subsurface ocean results form an interpretation of Galileo magnetometer data that requires an electrically conducting layer at a shallow depth in which a magnetic field is induced as Europa moves through the magnetosphere of Jupiter [404]. The observed magnetic field perturbations

Fig. 6.26. A cutaway view of the possible internal structure of Europa, showing a metallic iron, nickel core, surrounded by rock or silicate, with a thin outer layer of water in ice or liquid form, or perhaps a thin ice layer overlying a liquid water ocean. (Credit: NASA/JPL/Caltech)

are approximately those expected for moons responding as perfectly conducting spheres. Such a response requires a globally distributed highly conducting medium located close to the surface of the moon. This result has been interpreted as support for the presence of a salty sub-surface ocean. A subsurface ocean with the salinity typical of Earth's oceans and a few kilometers thickness can easily produce the observed induction response. Perhaps the most spectacular result of these observations is that the magnetic evidence indicates the existence of an ocean on Europa at this present epoch, not just during the recent past [405]!

The ice shell itself has been estimated by Shoemaker and others to be ~10 km thick. The ability of large craters to support central peaks, and the unusual morphologic transitions compared to craters on Ganymede and Callisto both suggest that the shell is greater than 5 km and could be as thick as 20 km according to Nimmo et al., [406]. Though the thickness of the ice remains controversial, it is clear that it has been disrupted extensively from below [407].

The discovery by the Galileo magnetometer team of what appeared to be an induced magnetic field with certain properties at Europa suggests that a large fraction of the water layer is a briny, liquid ocean ~100 km deep. If the ocean lacked appreciable solutes, or if it was thinner than a few tens of kilometers or frozen completely, it would not conduct current to explain the observed magnetic signature. Thus, the evidence for a subsurface ocean is strong [408]. The water layer (ice shell and water ocean) is probably composed of three sub-layers: an outer, brittle/elastic ice layer, an underlying ductile layer of potentially convecting ice, and a lower layer of liquid. The densities of these layers are expected to be greater for the lower layers, although some low-density pockets (briny ice/water or temperature-driven density anomalies) may occur throughout [409].

Melosh et al., argue that most of Europa's ocean is at the temperature of maximum density and that the bulk of the vigorously convecting ocean is separated from the bottom of the ice shell by a thin "stratosphere" of stably stratified water which is at the freezing point, and as a result, buoyant. Europa's overall density indicates it is mainly composed of silicates, not ice, and therefore contains radioactive heat-producing elements. This is enough to keep water beneath the ice shell stirred by vigorous thermal convection [410].

There has been much speculation that Europa's ocean might harbor life or, at least, prebiotic conditions. According to Kargel et al., just as life on Earth requires liquid water, chemical disequilibria, and elemental building blocks, we may find that these same stipulations are met on Europa and may have existed continuously or sporadically since Europa's origin. Molecular and isotopic studies of gas hydrates in lightless chemosynthetic communities of the Gulf of Mexico provide evidence that bacteria can directly oxidize hydrate-bound methane. Hydrocarbons from decomposing gas hydrates appear to drive complex biogeochemical processes in sediments that surround gas hydrate extrusion features helping to support complex chemosynthetic microorganisms. Metabolic processes such as methanogenesis, sulfur reduction, and iron oxide reduction have been suggested for Europa. Excreted byproducts and decaying material from chemoautotrophic organisms could serve as the basis for a Europan ecosystem by supplying carbon and energy sources to heterotrophic life forms at higher trophic levels. Photosynthesis would be unnecessary in this scenario. According to some geologic interpretations of Europa, there could in fact be enough chemical energy available on Europa to drive the development of a biomass. Calculations indicate that there may also be enough organic carbon to allow the existence of a dense, well-developed ecosystem. Even in the unlikely absence of endogenic sources of organic carbon, impacts can deliver organics, including amino acids. While temperature can certainly play a part in the development of life in an ocean, the exact temperature of an ocean can depend upon pressure and solutes. And, at any given location, the local temperature could be warmer near a hydrothermal vent. Thus, there could be conditions in Europa's ocean similar to conditions on Earth. Where Europan and terrestrial environments overlap in their physiochemical nature, examples from Earth reveal that life can cope and grow under conditions as extreme as the range of conditions expected for Europa. It is certainly possible that there are conditions on Europa that are far colder and of higher pressure than on Earth, but it is not certain that either of these conditions would be prohibitive to life as long as liquid water is present to allow organisms to grow. On Earth hypersaline, acidic, alkaline, and metal-rich environments harbor tenacious microbial communities. Therefore, it appears that organisms are able to exist and thrive under an extraordinary range of conditions wherever fresh liquid water or brines are present. Likewise, brine-filled cracks in sea ice or in the lithosphere provide a suitable habitat for microorganisms. Because life survives and grows in all of the coldest environments on Earth where liquid is present, even colder conditions may allow biological growth as long as liquid water is stabilized by solutes. Environments of even more extreme conditions may be surmountable if such conditions were stable. The adaptive strategies of microorganisms on Earth reveal that most every physiological stress can be overcome so long as the environment contains liquid water. Thus, it is possible that these types of environments could offer hope for life in the ice and rocky interior of Europa [411].

Seeming to agree with Kargel et al., Lipps and Rieboldt assert that Europa may harbor life because of the presence of a briny ocean, energy sources, and nutrient supplies. And, they point out that Europa's ocean probably contains far more water than oceans on Earth [412].

While some scientists continue to consider the existence of an ocean on Europa to be debatable, the interpretation by some scientists of the Galileo experiment using its magnetometer proved almost beyond doubt that Europa does indeed have an ocean beneath its frozen surface! Does this ocean hold some form of life? Only time and future exploration can tell us that. Perhaps a future orbiter and lander will give us the data we need [413, 414].

6.1.3 Ganymede

One astronaut who landed on the Moon described the terrain he saw there as 'Magnificent desolation!' Scientists, first seeing Ganymede's surface in detail, could easily have used the same description. Ganymede's surface is a tangle of geographic features! It has been pushed, pulled, and racked with unimaginable tectonic stress, bombarded with impacts, and then frozen in place (Fig. 6.27).

The surface temperature on Ganymede is 132–143 K [415], it is 5,268 km in diameter with a mean density of $1.936 \pm 0.022\,\mathrm{g\,cm^{-3}}$ [416], and completes an orbit of Jupiter in 7.155 days [417]. Ganymede is the largest satellite in the solar system, even larger than the planet Mercury.

Low resolution views of Ganymede obtained with Voyager show a surface divided into dark regions that are heavily cratered, and lighter regions organized into massive bands of grooved terrain that is much less heavily cratered. Higher

Fig. 6.27. A Galileo spacecraft view of Ganymede in natural color. The dark areas are the older, more heavily cratered regions and the light areas are younger tectonically deformed regions. The brownish-gray color is due to mixtures of rocky materials and ice. Bright spots are geologically recent impact craters and their ejecta. North is to the *top* of the picture. (Credit: NASA/JPL/Caltech)

resolution images, particularly those from the Galileo spacecraft, reveal craters with bright and dark ejecta in the darker regions (Fig. 6.28).

The lighter grooved terrain is found to have parallel alternating grooves and ridges running for hundreds of kilometers. The troughs and crests of these are separated by 5–10 km horizontally and several hundred meters vertically. The grooves and ridges appear to be bundled into formations referred to as 'lineations'. These lineations can be seen in some places to cut across one another, suggesting a very complex tectonic process (Fig. 6.29). This terrain can include small hills. In certain areas there is smooth terrain that might be the result of melting of the grooved terrain from below, or the result of cryovolcanism. Upon close examination it is easy to see that the grooved terrain cuts into the dark cratered terrain, indicating that the dark terrain is older. A good example of Ganymede's dark terrain is a region named *Galileo Regio* (Fig. 6.30). *Galileo Regio* is a large oval area ~2,800–3,200 km in diameter [418]. This feature had been seen by early telescopic observers; Dollfus and others making sketches of albedo markings. More recently even amateur astronomers, with modern 10-in. telescopes utilizing CCD cameras and webcams, are capturing these albedo features!

While the light, grooved terrain certainly presents evidence of tectonism, Galileo discovered the affects of tectonic activity in the darker regions as well. The dark terrain of *Nicholson Regio* is intensely fractured by tectonic forces, and one large crater was seen to have been torn apart [419].

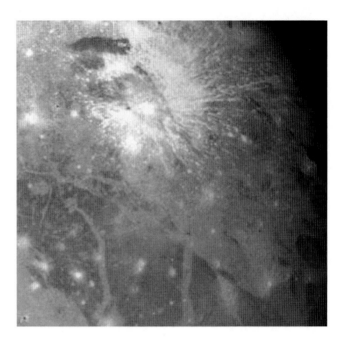

Fig. 6.28. A Galileo spacecraft view of the bright, rayed crater Osiris on Ganymede. North is to the *bottom* of the image. (Credit: NASA/JPL/Caltech)

Another dark region is *Marius Regio*. Marius Regio is a large area that lies near *Galileo Regio*. This combined area is of great interest because the two *regios* are separated by a *sulci* named *Uruk Sulcus* (Fig. 6.31). The separation of the two *regios* appears to have been the result of some form of plate tectonics [420]. Crustal spreading may be responsible for the formation of *Uruk Sulcus* [421].

About 65% of the surface area of Ganymede consists of bright terrain with relatively low crater populations. Most of this bright terrain is heavily grooved. These tectonically resurfaced areas generally show numerous triangular ridges and troughs, probably resulting from tilt-block normal faulting, resulting in the destruction of pre-existing surface forms. The eccentricity of Ganymede's early orbit may have been high enough in the past that tidal heating and flexing drove internal activity such that the internal heating aided the formation of bright terrain [422]. Many of the grooves probably formed at different times, during different deformational episodes, responding to different stress fields [423]. Pappalardo et al., made an initial analysis of Galileo images taken of the *Uruk Sulcus* area. Ridges and grooves are widespread in this region, with the terrain described as 'parallel ridged terrain'. The tectonic activity in the region is multifaceted. In addition to tilt block normal faulting, evidence is also found for horst and graben faulting, strike-slip deformation, high extensional strain, domino-style normal faulting, and horizontal shear and transtension. In *Uruk Sulcus*, spacing of ridges and troughs is ~8 km, with crest-to-trough height differences usually 300–400 m and as great as 700 m, with terracing apparent on some of the ridge walls [424]. Pappalardo et al., conclude there is abundant evidence that tilt-block-style normal faulting, horst-and-graben style normal faulting, and strike-slip deformation has modified pre-existing terrain through destruction of the older surfaces [425].

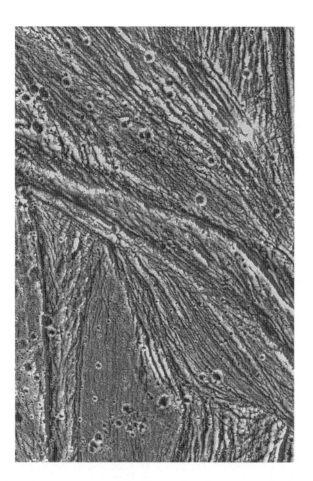

Fig. 6.29. Ridges, grooves, craters, and the relatively smooth areas in Uruk Sulcus on Ganymede were captured in this image by the Galileo spacecraft. The patterns of ridges and grooves indicate that extension (pulling apart) and shear (horizontal sliding) have both shaped the icy landscape. (Credit: NASA/JPL/Caltech)

The dark regions make up ~35% of the surface of Ganymede. Its geologic history is different from that of the bright areas. As alluded to, the dark terrain is covered with widespread, heavy cratering (Fig. 6.32). It is also faulted, to some extent. In the dark regions we see cratering, hummocky-hilly terrain, palimpsests, furrows, faults and fractures, knobs and massifs, furrow rims, and low-albedo plains. The geologic processes thought to have occurred in *Galileo Regio* in particular include tectonic deformation, mass wasting, sublimation, resurfacing by impact ejecta, and possible cryovolcanism and isostatic adjustment. However, scientists have not found unequivocal evidence for cryovolcanism within the dark regions; no features that may be source vents have been found. Impact cratering is a significant process within the dark regions of Ganymede. *Galileo Regio* is the largest contiguous region of dark terrain on Ganymede [426]. It is thought that the darker coloration of these regions consists of a thin, low albedo veneer overlying a

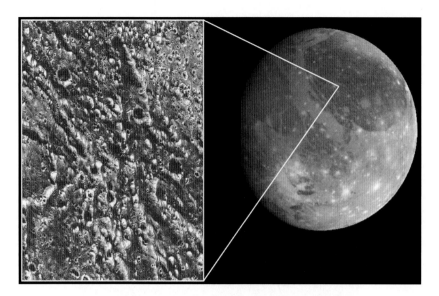

Fig. 6.30. This region of Galileo Regio contains ancient impact craters revealing the great age of the terrain. Galileo spacecraft images. (Credit: NASA/JPL/Caltech)

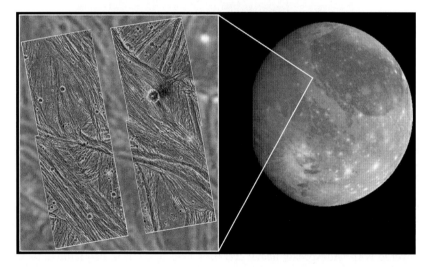

Fig. 6.31. A high-resolution mosaic of the Uruk Sulcus region on Ganymede, taken by the Galileo spacecraft. Note the parallel ridges and troughs. (Credit: NASA/JPL/Caltech)

cleaner substrate that contains small amounts of admixed dark meteoritic material [427]. By contrast, the low-lying ice of *sulci* is relatively clean.

There is significant impact cratering on Ganymede, indicating that Ganymede suffered its fair share of the effects of the 'Great Bombardment' period (Fig. 6.33). Most of the impacts appear to have occurred early in Ganymede's history. This is evident from the fact that most of the cratering evidence is contained in the dark

Fig. 6.32. A portion of the Galileo Regio region on Ganymede, revealing the heavily cratered dark terrain that makes up about half of the surface of this moon. Some craters are cut by numerous fractures, showing that the ancient crust was highly deformed early in Ganymede's history. A Galileo spacecraft image. (Credit: NASA/JPL/Caltech)

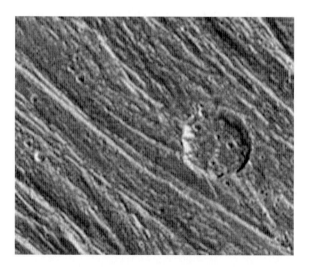

Fig. 6.33. The Galileo spacecraft reveals that the grooved terrain of Ganymede's Nippur Sulcus is composed of ridges and troughs spaced 1–2 km apart. A few broad ridges have smaller ridges on top of them. A 12 km crater is superimposed on these ridges. (Credit: NASA/JPL/Brown University)

regions of the surface, with the grooved terrain having resurfaced the moon in a later period. Yet, there are a number of impact crater features in these grooved areas, indicating that some of the impacts occurred after the grooves appeared. Relatively speaking, Ganymede is much more heavily cratered than Europa, but not nearly so as Callisto.

Craters on Ganymede do not behave the same as craters on our Moon or on Mars. For example, many of the craters on the Moon, large and small, show deep excavations with crater rims and central peaks. On Ganymede, as craters increase in diameter, the floors become shallower, and the crater rims less prominent; in other words, they become flatter with size. There are many examples of craters on Ganymede that are so shallow as to almost disappear. These features are referred to as 'palimpsests' and some of them are 400 km in diameter (Fig. 6.34). Palimpsests consist of four surface units: central plains, an unoriented massif facie, a concentric massif facie, and outer deposits. Deposits represent fluidized impact ejecta [428]. 'Palimpsests' account for a large portion of the surface and thus must have originated in Ganymede's early history. During this time Ganymede's surface was probably not as firm as it is now, and would have been able to flow over time and soften the relief of the crater. Or, alternately, the larger impacts were able to penetrate to the slush below at the time, allowing fluid to flood the crater floor.

There are young craters with brilliant ejecta ray systems that occur in both the dark regions and the light, grooved terrain (Fig. 6.35). The most conspicuous of these is named *Osiris*. This crater is 150 km in diameter with rays extending 1,000+ km [429]. Like our Moon, there is also evidence of some large impacts on Ganymede. There are several relics of multiple-ringed basins on Ganymede, with the largest of these being *Gilgamesh* in the southern hemisphere. *Gilgamesh* has a 150 km wide central depression, and an outermost ring with a 225 km radius [430].

The Galileo spacecraft detected a thin ionosphere, suggesting that Ganymede had a tenuous atmosphere. The Hubble Space Telescope had detected a tenuous oxygen atmosphere. Galileo also detected the presence of a magnetic field around Ganymede. This was also evidence for conditions allowing polar-aurorae. As it turns out, the prerequisites for an auroral display are a magnetic field, circulating charged particles, and a tenuous atmosphere. Ganymede has about 50% water ice on its surface. Apparently, both Europa and Ganymede release oxygen by

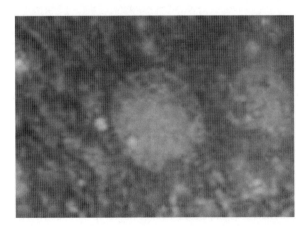

Fig. 6.34. A small portion of *Galileo Regio* showing the light-colored palimpsest of Memphis Facula. Although smooth in appearance, surprisingly the 340-km wide feature originated as the site of a massive impact. This Galileo spacecraft image reveals that the crater walls have slumped, the floor has risen isostatically, and the remaining topography has been smoothed out by slush. (Credit: NASA/JPL/Caltech)

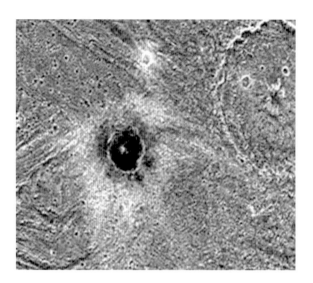

Fig. 6.35. The Galileo spacecraft caught this image of the dark-floored crater Khensu on Ganymede. Khensu possesses an unusually dark floor and a bright ejecta blanket. The dark component may be residual material from the impactor, or it may be that the impactor punched through the bright surface to reveal a dark layer beneath. (Credit: NASA/JPL/Brown University)

dissociation of water ice exposed on the surface. Hydrogen atoms are knocked off the surface by bombardment from charged particles. This process is called 'sputtering'. Because the hydrogen is lighter than oxygen, it leaks to space, leaving the oxygen behind. This type of atmosphere is very, very thin [431].

There is no sign of current cryovolcanism on Ganymede's surface [432]. In fact, the observations by the Galileo spacecraft provide little evidence for cryovolcanism on Ganymede; yet, Galileo did make a more detailed investigation of a suspicious region first seen by Voyager. Galileo confirmed the appearance of a feature in *Sippar Sulcus*, indicating a series of fluid eruptions creating a flow that appears to have eroded into the icy surface creating its own 'caldera' [433] (Fig. 6.36). So, while a few isolated cryovolcanic flow features have been discovered, there is little evidence for widespread volcanic landforms on Ganymede [434].

Ganymede is strongly differentiated. Previous modeling supports the hypothesis that Ganymede has an outer shell composed of various ice solid phases and a massive metallic core in which a magnetic field might be generated (Fig. 6.37). As a result of the Galileo spacecraft mission, the magnetic field has been confirmed and Ganymede is the first satellite found to have its own magnetosphere [435]. Not only does Ganymede posses its own intrinsic magnetic field (not an induced one), it also has its own magnetosphere, one that deflects Jupiter's magnetosphere! The core size of Ganymede ranges from 25 to 33% of its surface radius. The thickness of the silicate mantle above the core varies between 900 and 1,100 km [436]. Spectroscopic data suggests that ice is the major component on Ganymede's surface. Its mean density suggests that Ganymede's interior consists of approximately 60% silicate rock plus metal and 40% volatile ices by mass.

Its moment of inertia factor is the smallest measured value for any solid body in the Solar System and indicates a strong concentration of mass towards its center

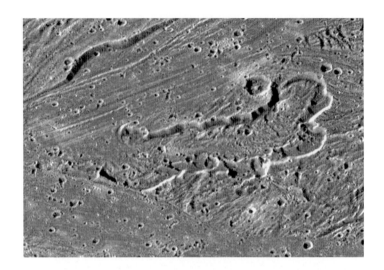

Fig. 6.36. The Sippar Sulcus area on Ganymede contains curvilinear and arcuate scarps or cliffs. These features appear to be depressions that might be sources for water ice volcanism thought to form the bright grooved terrain on Ganymede. This structure is the best candidate seen for an icy volcanic lava flow on Ganymede. The morphology of this structure suggests the possibility of volcanic eruptions creating a channel and flow, and cutting down into the surface. (Credit: NASA/JPL/Brown University)

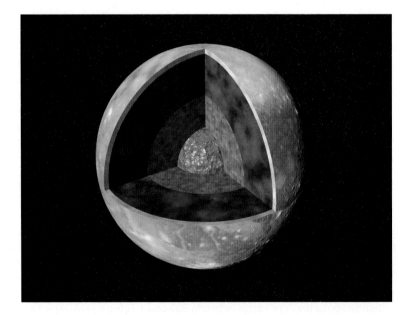

Fig. 6.37. This cut-out model represents the probable internal structure of Ganymede. Ganymede's surface is rich in water ice. The Galileo spacecraft confirmed that the moon is highly differentiated, with a rock and iron core overlain by a deep layer of warm soft ice, capped by a thin cold rigid ice crust. Data suggests that a dense metallic iron core exists at the center of the rock core. (Credit: NASA/JPL/Caltech)

[437]. According to John Anderson of JPL, 'the Galileo data showed clearly that Ganymede was differentiated into a core and a mantle, that there was a 800 km thick layer of warm ice beneath a warped and faulted ice crust, an equally thick mantle of rock, and an iron core.' It was not clear whether the metallic core was pure iron or a mixture of iron and iron sulfide [438].

There is indirect geological and geophysical evidence that Ganymede may posses a subsurface salty liquid water ocean [439]. Ganymede possesses an intrinsic magnetic field and the most likely source is dynamo action in a liquid Fe–FeS core. It has therefore been concluded that Ganymede's interior should have an iron-rich core surrounded by a silicate rock mantle and by an outer shell of ice. The ice shell is suggested to be ~800 km thick and the core may have a radius of between 400 and 1,300 km. Interpretations of magnetic data from Galileo passes of Ganymede have suggested the presence of a conducting layer at a depth between 170 and 460 km in which a magnetic field is being induced. This suggests that Ganymede, just like Europa and Callisto, may have a subsurface ocean. However, it must be noted that the magnetic data can also be fitted with models that do not require a subsurface ocean [440]. The melting temperature of ice will be significantly reduced by small amounts of salts and/or incorporated volatiles such as methane and ammonia. If these elements are present, they could contribute to the liquidity of the ice, further enhancing the chance of Ganymede currently having a liquid ocean below its surface [441].

6.1.4 Callisto

Being the Galilean moon furthest from Jupiter, Callisto completes an orbit about its parent in 16.689 days [442]. It has a surface temperature of 142–157 K [443], a diameter of $4,820.6 \pm 3.0$ km with no detectable deviation from sphericity, and a density of $(1,834.4 \pm 3.4)$ kg m^{-3} [444]. It is the second largest Galilean moon. Callisto has the lowest surface brightness of the four Galilean moons, although it is brighter than Earth's Moon (Fig. 6.38).

Spectral data clearly show that the surface of Callisto is covered by water ice. Its surface is evenly distributed dark, heavily cratered terrain, lacking volcanic or tectonic landforms. Callisto's low albedo also indicates the presence of non-icy components on its surface [445]. Of interest, small craters less than 3 km in diameter occur less frequently on Callisto than on Ganymede (Fig. 6.39). This suggests the presence of an erosional process at scales smaller than 1 km such as sublimation of surface ices, CO_2 degassing, or the existence of more volatile ammonia ice [446]. The surface of Callisto is the oldest among the Galilean satellites, and its largest craters are hundreds of kilometers in diameter. Its main features were probably formed during heavy bombardment [447].

Magnetic field data from the Galileo spacecraft suggests that Callisto has a magnetic field that may be induced by interaction with Jupiter's magnetic field. An internal conducting layer, such as a subsurface ocean of at least 10 km thickness with the salinity of terrestrial saltwater, could explain this. If it exists, the subsurface ocean is probably at great depth since there is no outward sign of any internal activity affecting the surface [448]. The observed magnetic field perturbations are approximately those expected for moons responding as perfectly conducting spheres. Such a response requires a globally distributed highly conducting medium located close to the surface of the moon. This result has been interpreted as support for the

Fig. 6.38. An image of Callisto by the Voyager 2 spacecraft. The surface of Callisto is the most heavily cratered of the Galilean moons. The bright areas are ejecta thrown out from relatively young impact craters. (Credit: NASA/JPL/Caltech)

presence of a salty subsurface ocean. A subsurface ocean with the salinity typical of Earth's oceans and a few kilometers thickness can easily produce the observed induction response. Perhaps the most spectacular result of these observations is that the magnetic evidence indicates the existence of an ocean on Callisto at this present epoch, not just during the recent past [449]! Modeling also supports the hypothesis that Callisto may have an internal liquid water ocean [450].

The surface of Callisto is very old and shows no evidence of extensive resurfacing. Unlike Ganymede, Callisto does not show areas of grooved terrain, probably due to its undifferentiated state and the lack of significant tidal stress from Jupiter [451]. The surface of Callisto is heavily cratered, more so than Ganymede, with craters and more craters everywhere we look (Fig. 6.40). Some scientists have characterized its surface as boring, especially compared to the other Galilean moons, and its craters show very little evidence of any kind of significant erosion that would be caused by tectonic activity or cryovolcansim.

Unlike the other Galilean moons, there is no convincing evidence of volcanism of any kind. This lack of erosion can possibly be attributed to the fact that Calliso is at a greater distance from Jupiter than the other three moons, and therefore does not suffer from tidal heating, nor from the orbital resonances that the other moons impose on each other [452]. The monotony of its surface is broken by a fair number of large, multiple-ringed, impact basins. This indicates that Callisto suffered from asteroid related bombardment [453] (Fig. 6.41). The largest and most outstanding feature on its surface is the *Valhalla* multiple-ringed impact basin. *Valhalla* is comprised of a smooth central plain ~600 km in diameter and a series of rings extending to a radius of ~2,000 km [454]!

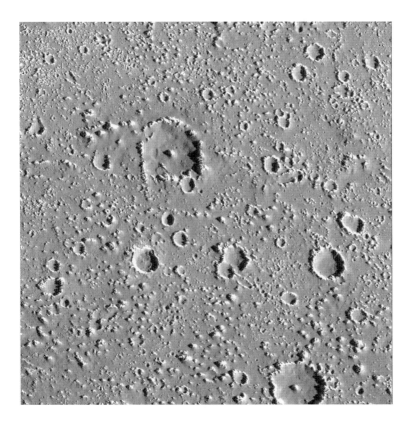

Fig. 6.39. Craters on Callisto as imaged by the Galileo spacecraft. Notice the dark, mobile blanket of dust-like material that seems to cover everything on Callisto. Some crater walls show movement of this material. While the moon has a significant number of large craters, it seems to lack a related number of small craters. While the dust-like material would fill in some small craters on the slopes of larger ones, it is not clear what process would erase the others. (Credit: NASA/JPL/Caltech)

In many places, the surface of Callisto gives the appearance of having been dusted in a dark powder. Whatever this dusty material is, it has the effect of softening the features of many of the craters on Callisto's surface [455]. Although not extensively resurfaced, there is some evidence of degradation of many of Callisto's craters. Landslides, slumping, and crater rim failure is seen in many areas [456, 457] (Fig. 6.42). There are some small 'smooth' areas suggestive of cryovolcanic resurfacing; however, there is no definitive evidence of this. There is quite a lot of evidence of other processes that have degraded Callisto's surface, such as sublimation-driven landform modification and mass wasting or slumping [458].

Like Ganymede, 'palimpsests' are evident on the surface of Callisto. Palimpsests are large, circular, low relief impact scars, and consist of four surface units: central plains, unoriented massif facies, concentric massif facies, and outer deposits. Palimpsest deposits represent fluidized impact ejecta rather than cryovolcanic deposits or ancient crater interiors [459]. Palimpsests give the impression of large craters that are almost not there.

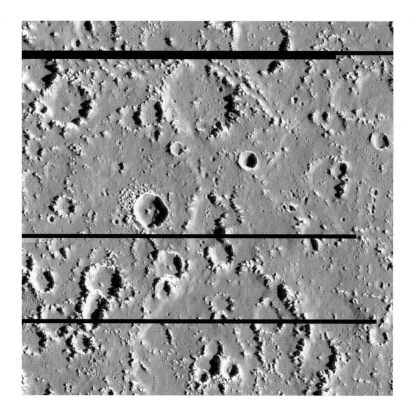

Fig. 6.40. Another portion of Callisto's surface imaged by the Galileo spacecraft. Note the varied landscape and the many features that are almost completely filled in by the dust-like material on the moon's surface. (Credit: NASA/JPL/Caltech)

Callisto possesses an exosphere. The ultraviolet spectrometer on board Galileo detected hydrogen atoms escaping from Callisto, and scientists determined that oxygen was therefore being disassociated from water molecules in the crust as a result of bombardment from solar ultraviolet radiation! The infrared spectrometer also found evidence of carbon dioxide frost and even gaseous carbon dioxide. According to Robert Carlson, Callisto's atmosphere is so tenuous that the carbon dioxide molecules drift around without bumping into one another. Callisto is not able to retain this atmosphere because ultraviolet radiation from the Sun breaks the molecules down into ions and electrons that are then swept away by Jupiter's magnetic field. It may be that carbon dioxide is periodically vented from the surface [460]. Otherwise, Callisto has no significant atmosphere.

There is indirect geological and geophysical evidence that Callisto may posses a subsurface salty liquid water ocean [461]. Callisto possesses one of the oldest known surfaces in the Solar system. Being the furthest from Jupiter of the Galilean moons, Callisto is not tortured by the tidal forces that can contribute to constant upheaval and resurfacing, and therefore avoided an internal heating process. The moment of inertia determined by the Galileo spacecraft suggests that Callisto has only a partially differentiated interior (Fig. 6.43). That is, it does not posses a distinct

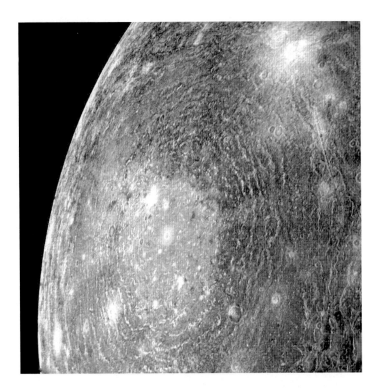

Fig. 6.41. A Voyager spacecraft image of the Valhalla multiple-ringed structure on Callisto. This feature consists of a light floored central basin some 300 km in diameter surrounded by at least eight discontinuous rhythmically spaced ridges. The rings are indicative of the moon's low density and probable low internal strength. (Credit: NASA/JPL/Caltech)

separate core and separate mantle, but ice and rock incompletely separated. Callisto is believed to have avoided differentiation through large-scale melting but may be incompletely differentiated through the convective gradual unmixing of two solid components, ice and metal-rich rock. It is not clear whether this unmixing process is still ongoing or has been arrested. Magnetic field data returned by Galileo suggests that Callisto has a conducting layer at a depth of not more than a few 100 km in which a magnetic field is induced. The simplest explanation for this observation calls for a subsurface ocean or layers of partially molten ice. Spectroscopic data suggests that ice is the major component on Callisto's surface. Its density suggests that Callisto contains approximately equal shares of rock/iron and ice [462, 463]. The source of heat that would keep a subsurface ocean from freezing completely on Callisto is probably radiogenic. Also, any salt in the ice or slush will act as a natural antifreeze. A salt mixture no more concentrated than the oceans on Earth would be enough to explain the magnetic field data collected by Galileo [464].

Kruskov and Kronrod suggested a six-layer, partially differentiated interior structure for Callisto, consisting of the following shells: (1) an outer water–ice envelope; (2) an intermediate water–ice mantle, which is subdivided into three reservoirs and composed of a mixture of high-pressure ices and rock material (dry silicates and/or hydrous silicates + Fe–FeS alloy); and (3) a central iron rock

Fig. 6.42. A Galileo spacecraft image revealing large landslide deposits within two large impact craters. The two landslides are about 3 km in length and occurred when material in the crater walls failed under gravity. These deposits are interesting because they traveled several kilometers from the crater wall in the absence of an atmosphere or other fluids that might have lubricated the flow. (Credit: NASA/JPL/Arizona State University)

ice-free core made up of a mixture of rock material and iron sulfide alloy [465]. Callisto does not display any non-impact related tectonic features [466].

So, of the four Galilean moons, Callisto has the oldest undisturbed surface and is the most stable. It may have a subsurface ocean that still exists today. And, unlike the other Galilean moons, it shows no sign of any kind of volcanism.

6.2 The Lesser Moons

Jupiter possesses 63 moons or satellites as of this writing, including a large number of smaller ones (Scott S. Sheppard, personal communication). Some of these small satellites reside in prograde orbits, and others reside in retrograde orbits (Table 6.1). Satellites that orbit in the same direction as the planet's rotation are *prograde*, and those that orbit opposite the direction of the planet's rotation are *retrograde*. Some

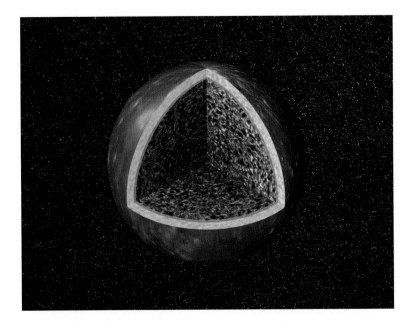

Fig. 6.43. An artist's concept of the interior of Callisto, depicting a salty ocean beneath an icy crust. Prior to Galileo, scientists thought Callisto was relatively inactive. Callisto's cratered surface lies at the top of an ice layer. Immediately beneath the ice, the blue band represents the possible ocean whose depth must exceed 10 km. The mottled interior is composed of ice and rock. (Credit: NASA/JPL/Caltech)

satellites have small, circular orbits with low inclinations, and the others have very large, very elongated orbits with high inclinations (Fig. 6.44).

6.2.1 The Regular Satellites

The category of *regular satellites* includes the Galilean moons previously discussed. The other regular satellites are very small bodies that have small circular orbits and low inclinations. Besides the four Galilean moons there are four inner regular satellites or moons. These are Metis 44 km, Adrastea 16 km, Amalthea 168 km, and Thebe 98 km (Fig. 6.45). All of the regular satellites probably formed in the early circumjovian disk of gas and dust around Jupiter during Jupiter's formation. All regular satellites have prograde orbits (Fig. 6.46).

6.2.2 The Irregular Satellites

There are a large number of *irregular satellites* orbiting Jupiter. Irregular satellites orbit Jupiter in very large orbits, with large inclinations and eccentricities, especially compared to the regular satellites [467] (Fig. 6.47). Some irregular satellites have prograde orbits and others have retrograde orbits. These bodies

Table 6.1. Jupiter satellite data. (Credit: Scott C. Sheppard)

Name/Designation		a (km)	I (°)	e	Period (days)	mag	H (mag)	Size (km)	Year
Small inner regulars									
XVI Metis		128,100	0.021	0.001	0.30	17.5	×	44	1979
XV Adrastea		128,900	0.027	0.002	0.30	18.7	×	16	1979
V Amalthea		181,400	0.389	0.003	0.50	14.1	×	168	1892
XIV Thebe		221,900	1.070	0.018	0.68	16.0	×	98	1979
Galileans									
I Io		421,800	0.036	0.000	1.77	5.0	×	3,643	1610
II Europa		671,100	0.467	0.000	3.55	5.3	×	3,122	1610
III Ganymede		1,070,400	0.172	0.001	7.16	4.6	×	5,262	1610
IV Callisto		1,882,700	0.307	0.007	16.69	5.7	×	4,821	1610
Themisto Prograde irregular									
XVIII Themisto	S/2000 J1	7,507,000	43.08	0.242	130.0	21.0	14.4	9	2000
Himalia Prograde irregulars									
XIII Leda		11,165,000	27.46	0.164	240.9	20.2	13.5	18	1974
VI Himalia		11,461,000	27.50	0.162	250.6	14.8	8.1	160	1904
X Lysithea		11,717,000	28.30	0.112	259.2	18.2	11.7	38	1938
VII Elara		11,741,000	26.63	0.217	259.6	16.6	10.0	78	1905
	S/2000 J11	12,555,000	28.30	0.248	287.0	22.4	16.1	4	2000
Carpo Prograde irregular									
XLVI Carpo	S/2003 J20	16,989,000	51.4	0.430	456.1	23.0	15.6	3	2003
Ananke Retrograde irregulars									
XXXIV Euporie	S/2001 J10	19,302,000	145.8	0.144	550.7	23.1	16.5	2	2001
XXXV Orthosie	S/2001 J9	20,721,000	145.9	0.281	622.6	23.1	16.5	2	2001
XXXIII Euanthe	S/2001 J7	20,799,000	148.9	0.232	620.6	22.8	16.2	3	2001
XXIX Thyone	S/2001 J2	20,940,000	148.5	0.229	627.3	22.3	15.7	4	2001
XL Mneme	S/2003 J21	21,069,000	148.6	0.227	620.0	23.3	16.3	2	2003
XXII Harpalyke	S/2000 J5	21,105,000	148.6	0.226	623.3	22.2	15.2	4	2000
XXX Hermippe	S/2001 J3	21,131,000	150.7	0.210	633.9	22.1	15.5	4	2001

XXVII	Praxidike	S/2000 J7	21,147,000	149.0	0.230	625.3	21.2	15.0	7	2000
XLII	Thelxinoe	S/2003 J22	21,162,000	151.4	0.221	628.1	23.5	16.4	2	2003
XXIV	Iocaste	S/2000 J3	21,269,000	149.4	0.216	631.5	21.8	14.5	5	2000
XII	Ananke		21,276,000	148.9	0.244	610.5	18.9	12.2	28	1951
Carme Retrograde irregulars										
XLIII	Arche	S/2002 J1	22,931,000	165.0	0.259	723.9	22.8	16.4	3	2002
XXXVIII	Pasithee	S/2001 J6	23,096,000	165.1	0.267	719.5	23.2	16.6	2	2001
XXI	Chaldene	S/2000 J10	23,179,000	165.2	0.251	723.8	22.5	15.7	4	2000
XXXVII	Kale	S/2001 J8	23,217,000	165.0	0.260	729.5	23.0	16.4	2	2001
XXVI	Isonoe	S/2000 J6	23,217,000	165.2	0.246	725.5	22.5	15.9	4	2000
XXXI	Aitne	S/2001 J11	23,231,000	165.1	0.264	730.2	22.7	16.1	3	2001
XXV	Erinome	S/2000 J4	23,279,000	164.9	0.266	728.3	22.8	16.0	3	2000
XX	Taygete	S/2000 J9	23,360,000	165.2	0.252	732.2	21.9	15.4	5	2000
XI	Carme		23,404,000	164.9	0.253	702.3	17.9	11.3	46	1938
XXIII	Kalyke	S/2000 J2	23,583,000	165.2	0.245	743.0	21.8	15.3	5	2000
XLVII	Eukelade	S/2003 J1	23,661,000	165.5	0.272	746.4	22.6	15.0	4	2003
XLIV	Kallichore	S/2003 J11	24,043,000	165.5	0.264	764.7	23.7	16.8	2	2003
Pasiphae Retrograde irregulars										
XLV	Helike	S/2003 J6	21,263,000	154.8	0.156	634.8	22.6	16.0	4	2003
XXXII	Eurydome	S/2001 J4	22,865,000	150.3	0.276	717.3	22.7	16.1	3	2001
XXVIII	Autonoe	S/2001 J1	23,039,000	152.9	0.334	762.7	22.0	15.4	4	2001
XXXVI	Sponde	S/2001 J5	23,487,000	151.0	0.312	748.3	23.0	16.4	2	2001
VIII	Pasiphae		23,624,000	151.4	0.409	708.0	16.9	10.3	58	1908
XIX	Megaclite	S/2000 J8	23,806,000	152.8	0.421	752.8	21.7	15.0	6	2000
IX	Sinope		23,939,000	158.1	0.250	724.5	18.3	11.6	38	1914
XXXIX	Hegemone	S/2000 J8	23,947,000	155.2	0.328	739.6	22.8	15.9	3	2003
XLI	Aoede	S/2003 J7	23,981,000	158.3	0.432	761.5	22.5	15.8	4	2003
XVII	Callirrhoe	S/1999 J1	24,102,000	147.1	0.283	758.8	20.8	14.2	7	1999
XLVIII	Cyllene	S/2003 J13	24,349,000	149.3	0.319	737.8	23.2	16.2	2	2003

(Continued)

Table 6.1. (continued)

Name/Designation	a (km)	I (°)	e	Period (days)	mag	H (mag)	Size (km)	Year
Jupiter satellites discovered in 2003 yet to be named								
S/2003 J2	28,570,410	151.8	0.380	982.5	23.2	16.6	2	2003
S/2003 J3	18,339,885	143.7	0.241	504.0	23.4	16.9	2	2003
S/2003 J4	23,257,920	144.9	0.204	723.2	23.0	16.4	2	2003
S/2003 J5	24,084,180	165.0	0.210	759.7	22.4	15.6	4	2003
S/2003 J9	22,441,680	164.5	0.269	683.0	23.7	17.2	1	2003
S/2003 J10	24,249,600	164.1	0.214	767.0	23.6	16.7	2	2003
S/2003 J12	19,002,480	145.8	0.376	533.3	23.9	17.2	1	2003
S/2003 J14	25,000,000	140.9	0.222	807.8	23.6	16.7	2	2003
S/2003 J15	22,000,000	140.8	0.110	668.4	23.5	16.8	2	2003
S/2003 J16	21,000,000	148.6	0.270	595.4	23.3	16.3	2	2003
S/2003 J17	22,000,000	163.7	0.190	690.3	23.4	16.5	2	2003
S/2003 J18	20,700,000	146.5	0.119	606.3	23.4	16.5	2	2003
S/2003 J19	22,800,000	162.9	0.334	701.3	23.7	16.7	2	2003
S/2003 J23	24,055,500	149.2	0.309	759.7	23.6	16.7	2	2003

A – The mean semi-major axis
I – The mean inclination
E – The mean eccentricity
Period – The time for one revolution around Jupiter
mag – The optical magnitude of the object (R-band)
H – The absolute magnitude of the object
Size – The diameter of the object
Year – The year of discovery
(Courtesy Scott S. Sheppard.)

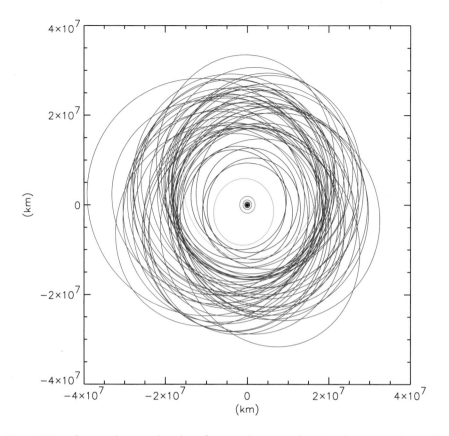

Fig. 6.44. A diagram depicting the orbits of Jupiter's known satellites. (Credit: Scott C. Sheppard)

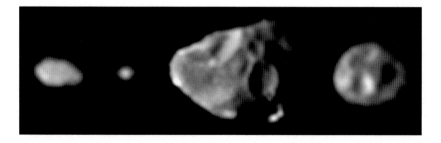

Fig. 6.45. Galileo spacecraft images showing the four small, irregular shaped moons that orbit Jupiter in the zone between the planet's ring and larger Galilean satellites. Shown in correct relative size and from *left* to *right*, arranged in order of increasing distance from Jupiter, are Metis, Adrastea, Amalthea, and Thebe. (Credit: NASA/JPL/Cornell University)

are very small. Because of their small size and orbital characteristics, they are believed to be objects that were captured from orbits around the sun during the early formation of Jupiter [468]. Himalia is the largest irregular satellite, having

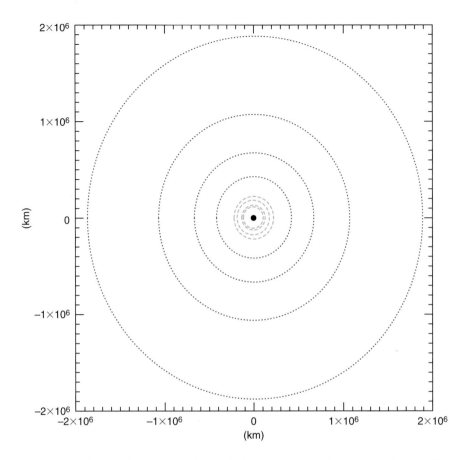

Fig. 6.46. A diagram depicting the orbits of the known Jovian regular satellites. These satellites have small, circular orbits and low inclinations. (Credit: Scott C. Sheppard)

a diameter ~150 km. This little moon was discovered by Perrine in 1904. There appear to be two distinct prograde groups of satellites and at least three retrograde groups. The groups suggest that these satellites are the result of the breakup from collision of multiple parent bodies. Breakup could be the result of impact with interplanetary bodies, primarily comets, or by collision with other satellites. Each retrograde group contains one large object (with a radius >14 km) along with several smaller ones (with a radius < 4 km). Outside of these distinct groups, there are many satellites with diameters of just 2 km. Recent surveys predict that Jupiter should be surrounded by ~100 rocky satellites with diameters larger than ~1 km [469].

Evidence from physical observations of these satellites is limited, but we know that their colors range from neutral, the color of the Sun, to moderately red. The satellites lack the ultra-red material found on the Centaurs and Kuiper Belt Objects. The retrograde satellites are systematically redder than the prograde satellites, suggesting that the retrograde satellites are fragments of a D-type body while the prograde satellites result from the breakup of a C-type body [470]. However, while

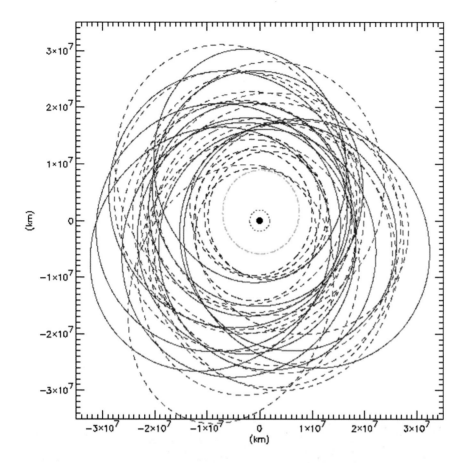

Fig. 6.47. A diagram depicting the orbits of the Jovian irregular satellites. Irregular satellites have large orbits, inclinations, and eccentricities. These are probably objects captured during the early formation of Jupiter. (Credit: Scott C. Sheppard)

the colors of irregular satellites are very similar to C, P, and D-type carbonaceous outer main belt asteroids, the spectra of Jupiter's irregular satellites are consistent with C-type asteroids [471]. Evidence for oxidized iron in phyllosilicates has been found in an absorption feature in the spectra of the moon Himalia [472].

While the regular satellites, with their circular, non-eccentric orbits are expected to exist in permanent orbits around Jupiter, the irregular satellites must be considered temporarily captured. In fact there are several examples of comets that are in temporary capture orbits around Jupiter, and others that scientists expect to be captured in the next 100 years. The most famous example of a temporarily captured satellite is comet Shoemaker-Levy 9 (SL9), which dramatically crashed into Jupiter in 1994. SL9 was probably in orbit around Jupiter for decades to a century. Had it not collided with Jupiter it would eventually have been ejected by the planet, either to orbit the Sun as a short period comet or to leave the solar system completely [473].

6.3 The Rings of Jupiter

I remember being fascinated as a young boy that the planet Saturn possessed a ring system. As I grew up through high school and even into college, that fascination remained with me. As I look back now, I realize it never occurred to me and perhaps to no one else, that we would one day use the phrase 'The Rings of Jupiter'! How times have changed. The discovery that other planets have rings should be the ultimate lesson for us in expecting the unexpected, especially in astronomy!

Jupiter was not the first planet after Saturn to enter the family of ringed planets. That honor actually fell to the planet Uranus when, on March 10, 1977 teams led by James L. Elliot and Robert L. Mills observed the stellar occultation of the star SAO 158687 by Uranus. An analysis of the premature, unexpected dip in brightness of the star led to the conclusion that Uranus possessed rings. Subsequent to this discovery, the Voyager 1 spacecraft confirmed the existence of rings around Jupiter in March 1979 [474]. The Uranus rings were imaged for the first time when Voyager 2 encountered the planet. Finally, in Voyager 2 images taken on August 11, 1989 ring arcs were seen orbiting the planet Neptune. Where previous occultation observations of Neptune had failed to be conclusive, Voyager images confirmed the existence of Neptune's rings [475]. Thus, all four gas giant planets were found to have rings, and each planet's rings were found to be different from that of the others.

Jupiter's rings are organized into three main components; a cloud-like *halo*, a *main ring*, and a faint, *gossamer ring* [476] (Fig. 6.48). Jupiter's rings are optically thin, and contain large numbers of dust-sized particles [477]. Because of the way

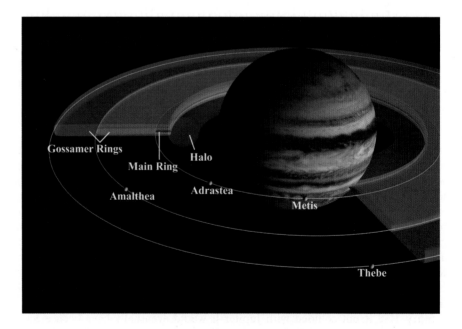

Fig. 6.48. A schematic cut-away view of the components of Jupiter's ring system showing the geometry of the rings in relation to Jupiter and the small inner satellites, which are the source of the dust that forms the rings. (Credit: NASA/JPL/Cornell University)

incident light is scattered by particles in the ring, the 'main' ring is thought to contain a significant population of micron-sized particles, ranging from micron-size to tens of micron-size [478]. Modeling suggests that ring particles drift inward towards Jupiter and are eventually lost to the planet's atmosphere [479]. Both gravitational and magnetic forces are thought to control the ring system [480].

Evidence of Jupiter's rings was first obtained by particle flux measurements made by the Pioneer 10 and 11 spacecraft. Conclusive proof was not obtained until the Voyager 1 spacecraft imaged the *main ring* in March 1979. Voyager 2 took additional images and revealed the hint of a broad ring outside of the main ring, now known as the *gossamer ring*. A wider, more diffuse *halo ring* was also detected orbiting interior to the main ring [481]. The Galileo spacecraft later confirmed the existence of the outer gossamer ring [482, 483]. Three small inner moons were also discovered during the Voyager encounter with Jupiter. These moons are called Thebes, Adrastea, and Metis [484, 485]. These inner moons, along with the inner moon Amalthea, not only supply the material for the rings, but also shepherd and help define the limits of the rings. The ring neighborhood where Jupiter's ring resides also contains several small lumpy satellites, or "collisional shards" as Burns refers to them [486].

The *halo ring* is the innermost component of the ring system and is a cloud of fine particles that bloom vertically at the main ring's inner boundary and continues with decreasing intensity toward the planet and vertically [487]. The halo ring extends from ~92,000 to ~122,500 km outward from Jupiter [488] (Fig. 6.49). The halo is caused by interaction with Jupiter's magnetic field, and is thought to be composed of fine dust grains that have been dragged inward from the main ring until they are excited vertically. It has a height of ~20,000 km [489–493].

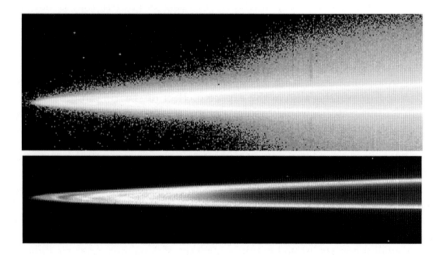

Fig. 6.49. Two images of Jupiter's rings, taken by the Galileo spacecraft. Different brightness scales accent different parts of the ring system. The ring system has three main parts – a flat main ring; a halo inside the main ring shaped like a double-convex lens; and the gossamer ring outside the main ring. In the top view, a faint mist of particles is seen above and below the main ring. This "halo" is unusual in planetary rings, and is caused by electromagnetic forces pushing the smallest grains, which carry electric charges, out of the ring plane. (Credit: NASA/JPL/Cornell University)

The *main ring* is the brightest and easiest ring to detect. It was seen clearly in Voyager images and later in ground based images [494]. According to Brooks et al., the main ring is thought to extend 6,500–7,000 km [495]. Ockert-Bell et al., suggest a width of ~6,440 km, spanning a distance of 122,500–128,940 km [496]. Cassini observations placed an upper limit on the rings thickness at 80 km [497]. The main ring possesses a sharp outer boundary and a diffuse inner one. The Galileo spacecraft also revealed that the ring displays a phenomenon known as the *Metis notch*, which is located near the outer boundary of the ring. The location of the Metis notch suggests a relationship with that moon, although the relationship is not fully understood. Perhaps it has something to do with the manner in which the moon shepherds material in the ring. The Metis notch appears as a significant decrease in brightness of the ring in the vicinity of Metis' orbit, which is bounded on its outer edge by a bright annulus of ring material. Galileo also found bright patches in the main ring whose origin is not understood. These bright patches may be debris generated by smaller, less energetic collisions [498, 499]. The Cassini spacecraft saw possible 1,000 km-scale azimuthal clumps within the ring, and ruled out the possibility that they could be previously undetected moons. A possible explanation is that they are simply slight density variations at that vicinity of the ring, a simple clumping of material. Or they may be no more than a natural consequence of a long line-of-sight [500]. The Galileo spacecraft also observed a diffuse and vertically extended cloud of material, reminiscent of the *halo*, above and below the main ring [501]. It is believed that particles for the main ring come from the small moons Adrastea and Metis, with Adrastea thought to be the primary source of material. Adrastea's orbit lies just outside the outer edge of the main ring. The orbit of Metis lies just 1,000 km inside Adrastea. Outside of Metis, the brightness of the main ring drops off very quickly.

There is some further asymmetry in the ring brightness that is not fully explained. Among the possible explanations for this is that asymmetry may arise from elongated ring particles oriented along a particular axis. Another is that it may be related to the magnetic field. Or, it may be due to localized enhancement in the number of ring particles. Another explanation may be the generation of debris from collisions or from an impact into a ring parent body by an external impactor. However, because of the expected effect of shearing in the rings, the origin for this material would require a more recent impact event and thus, this asymmetry might be a localized short term material-producing impact event. Light-scattering studies suggest that the ring parent-bodies are concentrated near the outer edge of the main ring, in the outer most ~2,000 km of the ring [502]. Parent bodies probably make up a majority of the ring system's mass, but a much smaller fraction of it surface area. Dynamical arguments suggest that parent-bodies are the collisional remnants of a fragmented satellite [503]. In November 2002, the Galileo spacecraft observed a series of flashes near the moon Amalthea. It is believed that these flashes represent sunlight reflected from 7 to 9 small moonlets located within ~3,000 km of Amalthea. Analysis indicates that these small bodies are between 0.5 m to several tens of km in diameter. In September 2003 Galileo detected a single additional body. It is suggested that these small bodies are a part of a discrete rocky ring embedded within Jupiter's gossamer ring [504].

The *gossamer ring* is the outer most member of the ring system. The gossamer ring has two components: one extends from the exterior of the main ring 128,940 km to just inside Amalthea's orbit at 181,366 km and the other extends

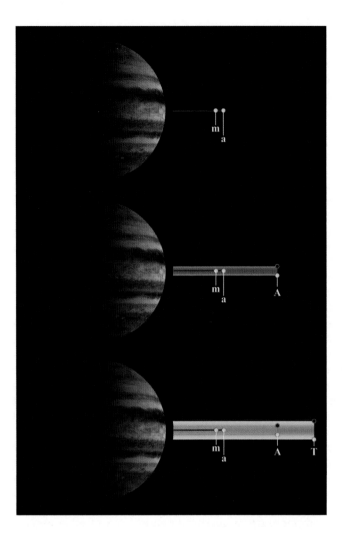

Fig. 6.50. A schematic depicting the structure of Jupiter's main and gossamer rings. The top panel shows that the main ring (red) is formed mostly from meteoroid impact debris kicked up from the innermost moons, Metis (m) and Adrastea (a). Since both satellites orbit in paths not inclined to Jupiter's equator, the main ring appears as a narrow line. The *Middle panel* shows the effect of dust ejected from the satellite Amalthea (A), responsible for producing one of the two moon components of the gossamer ring. Amalthea's orbit is inclined to Jupiter's plane, and at different times the satellite's vertical position can range anywhere between the two extreme limits shown. Dust ejected from Amalthea (orange) produces a ring whose thickness equals Amalthea's vertical positions beyond Jupiter's equatorial plane. The lower panel shows the additional effect of dust ejected from Thebe (T), which makes up the second component (*shown in green*) of the gossamer ring. The two positions shown represent the maximum projections of Thebe from Jupiter's equatorial plane. This component of the gossamer ring is thicker than the Amalthea dust component because Thebe's orbit is more inclined than that of Amalthea. (Credit: NASA/JPL/Cornell University)

from the exterior of the main ring out to just inside Thebes orbit at 221,888 km (Fig. 6.50). Very faint material can be detected out beyond the orbit of Thebes where it finally blends into the background at 250,000 km [505].

The moons Thebes and Amalthea are believed to be the source of the material in the translucent gossamer ring. These inner moons are heavily cratered by interplanetary meteor impacts and it is the dust from these impacts that sustain the rings. There are many clues that support this conclusion. Among them is the fact that the structure of the outer gossamer ring comprises two rings, one embedded within the other! The orbits of Amalthea and Thebes are slightly inclined to Jupiter's equatorial plane. The outer gossamer rings are also slightly inclined to Jupiter's equatorial plane and the two ring components are even tilted at an angle to one another!

Ockert-Bell et al., refer to the innermost gossamer ring as the *Amalthea ring*, and the other as the *Thebe ring*. The outermost material beyond Thebe is simply referred to as the *gossamer extension*. The thickness of the Amalthea ring is thought to be < 4,000 km, and the Thebe ring is thought to be > 8,000 km across vertically [506]. The thickness of these rings seems to be connected to their corresponding moon and the elevation each moon reaches above and below the planet's equatorial plane. Amalthea and Thebes exhibit epicyclic latitudinal oscillations that carry them above and below Jupiter's equatorial plane and, as material is lost from the surface of these moons, it is deposited in the gossamer rings [507].

In addition to observations by the Voyager, Galileo, and Cassini spacecraft, Earth based observations have been carried out by the Palomar 5-m telescope and the IRTF telescope, the Keck 10-m telescope, and the Hubble Space Telescope. These observations indicate that the material in the rings is very red in color. This red color comes from both the intrinsic color of the large bodies, and the red light preferentially scattered by the dust. The observations also indicate that the particles in the ring are both non-spherical small particles of sizes up to tens of microns and perhaps mm- to km-size large bodies. This non-spherical aspect fits with the effects of impacts and collisions between bodies [508].

The rings have been observed from Earth with professional size instruments in visible, near infrared, and infrared wavelengths. To my knowledge no amateur with amateur equipment has yet managed to image Jupiter's ring system. However, with the advancement in equipment and telescopes that are becoming available to amateurs, how long will it be before this happens? My own astronomical society possesses an advanced 0.6-m instrument and I know of other organizations that possess larger ones. Surely, this observation will not continue to be the domain of professionals forever.

6.4 Trojans and Comets

6.4.1 Trojans

Jupiter is such a massive planet that it controls its own family of asteroids. These bodies are known as *Trojans* and reside in Trojan clouds located at or near Jupiter's lagrangian points L4 and L5 (Fig. 6.51). The lagrangian points are locations along Jupiter's orbit where the gravitational attraction from the Sun and Jupiter are balanced. Trojans have diameters < 300 km, yet their numbers are so great that they number as many objects as contained in the main asteroid belt located between the orbits of Mars and Jupiter.

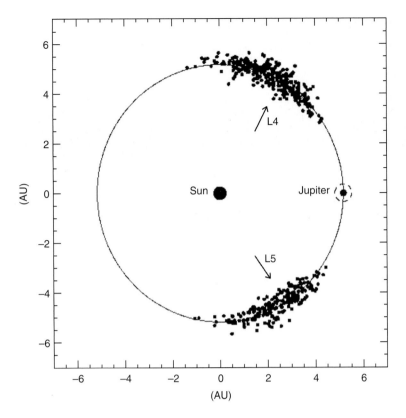

Fig. 6.51. A simplified view of the Trojan asteroid clouds at Jupiter's L4 and L5 Lagrangian points. These clouds are composed of small bodies with diameters < 300 km. The gravitational attraction from the Sun and Jupiter are balanced near these areas, allowing the objects to have stable orbits over the age of the Solar system. The *dashed circle* around Jupiter shows its Hill sphere, or the area where Jupiter's gravity dominates that of the Sun. This Hill sphere is the area in which all of Jupiter's satellites are found. (Credit: Scott C. Sheppard)

6.4.2 Jupiter Family Comets

Jupiter also has great influence over comets that fall in toward the Sun. Short period comets, having orbital periods < 20 years and low inclinations, are controlled by Jupiter. Short period comets are thought to originate from the Kuiper Belt. From time to time, the orbits of these objects may be jostled due to collisions or gravitational affects, altering their orbits and sending them on a new path in toward the Sun. The new orbit of the comet will cross the orbit of Jupiter, allowing gravitational interaction. Over time, the comet's orbit will gradually change until it is either thrown out of the solar system or it collides with a planet or the Sun. The most spectacular example of a planetary collision in our lifetime was that of comet Shoemaker-Levy 9, which crashed into Jupiter in 1994 (Fig. 6.52).

Fig. 6.52. A series of images showing the blemishes from the impact of Comet Shoemaker-Levy 9. These impact features were easily seen in amateur telescopes. (Credit: Donald Parker)

Section II

How to Observe the Planet Jupiter

Section II will discuss what truly excites us and gives us the greatest pleasure, the observation of the planet itself. We will discuss equipment, the types of observations, keeping a record, and reporting.

Introduction

Amateurs have contributed to the observational record of Jupiter for years and years. The official reports of the British Astronomical Association go back at least 114 years, to 1891. Individual reports by amateurs can be traced back even further than that to 1869, and Rogers (1995) shows sketches as far back as 1831. The Association of Lunar and Planetary Observers can claim organized records beginning in 1947, the year the organization was formed; and certainly individual A.L.P.O. members were making serious observations even before then. Organizations in Europe and the Orient have also made serious contributions to the observation of Jupiter. The amateur and professional Jupiter astronomers of today stand in awe of observers like T. E. R. Phillips, P. B. Molesworth, Antoniadi, F. J. Hargreaves, Bertrand M. Peek, Walter H. Haas, Elmer J. Reese, T. Sato, A. W. Heath, J. Dragresco, and many, many others. If the accomplishments of the scientists of the past are due in part because they stood on the shoulders of those who came before them, it is all the more true for those of us who observe and study Jupiter today. The truth is, without the observational records of amateurs, we would not have a continuous record of the physical appearance of Jupiter, since professional astronomers are often busy elsewhere. Even in this ultra-modern, high-tech world, amateur observations will continue be important to the study of this giant planet.

Since the professional community never seems to get enough observing time on the world's professional instruments, they need the observations of amateurs to fill the void in their own observations or even to be the basis of their research. Serious amateurs make careful, standardized observations that can be measured statistically. Professional astronomers need the data amateurs collect. We used to say that, although amateurs did not have equipment that was as sophisticated as the professional astronomer, the amateur's data, carefully recorded, is still valuable. Well, that statement is still true except for the part about sophistication. Larger, better telescopes are becoming more affordable and more sophisticated. Instrumentation, such as CCD cameras, are no longer only within reach of professionals, and even really large instruments are becoming available to amateurs as amateur organizations find ways to build sophisticated observatories.

So, what are the observations that amateurs can make? Actually, there are many types of observations that amateurs can perform. These observations include disk drawings, strip sketches, intensity estimates, central meridian transit timings and the construction of drift charts, measurements of latitude, observations of color, stellar occultations, eclipses and transits of the Galilean moons, CCD and webcam imaging, measurements of images, and photometry of Jupiter and its Galilean moons. To the beginner this list must seem unbelievable; can amateurs really do all of this? Yes, they can and we'll discuss each in turn. We'll also discuss the different types of telescopes, advantages or disadvantages of each type, eyepieces and filters, and other helpful pieces of equipment.

Chapter 7

Equipment

Any serious observing program will benefit from equipment that is up to the task. The quality of equipment used should be the best the observer can afford. For planetary work, some designs are better than others, and some should be avoided if at all possible. Here we will discuss what is available and the advantages and disadvantages of each choice.

7.1 Telescopes

To make meaningful observations of Jupiter, you will need access to an adequate telescope. There's the big question! What is an adequate telescope? What type of telescope is best? What is the minimum size telescope required? Many observers prefer refractors, others prefer Newtonian reflectors, many astronomers make meaningful observations with Schmidt-cassegrains, and some with exotic types. It is easy to understand why the beginner could be easily confused.

I believe that a refractor of at least 4-in. aperture or a reflector of at least 6-in. aperture is the minimum size needed to engage in serious observations of Jupiter. Telescopes of smaller size will not provide enough resolution. The term resolution refers to the ability of a telescope to resolve small details. In no other activity is it more important, that a good big one is better than a good little one. In planetary observing, size can give you an advantage. Normally, the larger the aperture of the telescope lens, the better its resolving power. The resolving power of a telescope can be stated in terms of the Dawes limit. The Dawes limit is expressed by the formula:

$$\text{sep}'' = \frac{4.56}{d},$$

where sep″ stands for the minimum separation in seconds of arc at which two stars of equal magnitude can be observed as two distinct objects separate from each other, in a telescope of a given aperture. The Dawes factor is 4.56, and d represents the diameter of the telescope lens or mirror in inches. So, the larger the telescope, the greater its ability to resolve two stars that are close together; the bigger the aperture, the smaller the separation that can be observed. Once, one of my professional astronomer friends informed me he had requested time on one of the Keck 10-m telescopes for a Jupiter observing run. The reason he wanted it was not its light gathering ability but its resolving power. Of course, resolution can be adversely affected by poor seeing and by optics of poor quality. Seldom will any telescope achieve the theoretical limit of resolution (Table 7.1). Some nights the seeing can be so poor that no telescope, regardless the aperture, would produce acceptable results. But on those nights when seeing is great, a larger telescope will gather more light, allow more magnification, and show more fine detail.

The contrast of features on Jupiter is very subtle; therefore, the desirable telescope for observing Jupiter is one that forms a sharp image with high image contrast. Most astronomers would argue that of all types, the refracting telescope does this best. Refractors make an image by focusing light through a clear lens. The lens is mounted in a lens cell at the front of a tube and light is brought to a focus at the lower end of the tube where an eyepiece or other instrument is attached. A refracting telescope that is well made is a dependable instrument (Fig. 7.1) requiring relatively little maintenance compared to reflecting telescopes. Refractors provide high image contrast since there is no obstruction in the light path as with traditional reflecting telescopes. Observers using high quality refractors have made some of the best planetary drawings I have seen.

However, refractors of large aperture are expensive to make. The objective lenses of early refractors were plagued with a certain amount of chromatic aberration. Because of this chromatic aberration, red and blue fringing was often seen around the edges of bright objects. This caused the color rendition of planets and bright stars to be somewhat unreliable in refractors. In years past, refracting lenses had to be constructed with long focal lengths to eliminate as much chromatic aberration as possible. This resulted in refracting telescopes with focal ratios of $f/15$ or higher. Really large refractors had optical tubes of great length and weight, requiring large mounts to support them and large structures to house them. Consequently, the 40-in. Yerkes refractor was the largest and the last of the large refractors to be made. Amateurs generally do not have such large instruments; but a 6-in. $f/15$ refractor still requires a lot of space to use and is not so easily transported.

Recently, especially in the last few years, manufacturers have overcome most of the problems inherent in refracting lenses. Modern high quality refractors use lenses made of special glass that are carefully ground by computerized machines to shorter focal lengths. Modern lenses are able to accurately focus all the different

Table 7.1. Resolution of telescopes according to Dawes' limit	
Aperture	Resolution (seconds of arc)
6 cm (2.4 in.)	2.2
8 cm (3.1 in.)	1.5
10 cm (12.0 in.)	1.2
15 cm (6.0 in.)	0.8
20.3 cm (8.0 in.)	0.6
25 cm (10.0 in.)	0.5
32 cm (12.5 in.)	0.4
40 cm (16.0 in.)	0.3
(After Sherrod 1981 [510].)	

Fig. 7.1. A typical refracting telescope on a German-equatorial mount. (Credit: Brady Richardson).

wavelengths of light into crisp, color accurate images. These developments have made the refracting telescope more affordable and more compact. Today, many observers are selecting refracting telescopes as their telescope of choice.

Next to the refractor, a long-focus Newtonian reflecting telescope (Fig. 7.2) produces the best image contrast. To many observers, a 6-in. f/8 Newtonian reflector offers as much image contrast as a 4-in. f/15 refractor. These reflectors use a parabolic shaped mirror, located at the lower end of the tube, to collect and bring light to a focus near the upper end of the optical tube. Because the glass for a mirror does not have to be as optically pure as a refracting lens, mirrors are easier and less expensive to make. Also, mirrors can be more easily ground to shorter focal lengths. As a result, the cost per inch of aperture is usually much lower for a reflecting telescope than for a refractor. And, because mirrors are much easier to grind accurately than refracting lenses, the mirror can be formed into a perfect parabolic shape so that every wavelength of light can be brought to the same focal point. This virtually eliminates chromatic aberration in a reflecting telescope. With shorter focal lengths, reflectors can be made shorter and lighter, the mounts do not have to be as large as refractors, and reflectors can be housed in smaller shelters. Consequently, even modestly large reflectors can be transportable. One drawback of reflecting telescopes is that they require more maintenance than refractors. The mirror for a reflecting telescope is mounted in a mirror cell that is itself mounted to the optical tube. To bring the focused image to an eyepiece, a secondary optical flat mirror set at a 45° angle is used at the upper end of the optical tube to divert the light path out the side of the tube to an eyepiece or other instrument. The two mirrors in this system must be perfectly collimated and aligned to produce a usable image. Periodically, these mirrors will require adjustment to keep the telescope in good working order. Also, the mirror

Fig. 7.2. A 12-in. (305 mm) Newtonian reflecting telescope on a German-equatorial mount. (Credit: Dave Eisfeldt/the Central Texas Astronomical Society).

surfaces will need to be cleaned and recoated with reflecting material from time to time. This takes time and money. Because the contrast of Jupiter's features is so subtle, slight misalignment of the optics will smear out fine detail that would otherwise be seen. In spite of these detractors, large reflectors are much more common than large refractors. Today's large, modern professional telescopes are all reflecting telescopes. Newtonian reflecting telescopes are great telescopes for observing Jupiter.

A variation on the Newtonian reflecting telescope is the classic cassegrain telescope (Fig. 7.3). Cassegrain telescopes also use a parabolic concave primary mirror but use a relatively small convex secondary mirror. Instead of reflecting the light out the side, light is reflected back down the tube through a small opening in the center of the primary mirror, arriving at focus behind the primary mirror. Cassegrain telescopes often have very long effective focal lengths, contributing to high image contrast and allowing high magnification. Many modern professional telescopes use the classic cassegrain configuration.

Today, more amateurs are using Schmidt-cassegrain telescopes (SCTs) than any other design. SCTs use a circular rather than parabolic mirror as the primary mirror. This primary mirror is located at the bottom end of the optical tube, as in a classic cassegrain telescope. Because a circular mirror is prone to chromatic aberration, a very thin correcting lens is placed at the front of the optical tube. Mounted on the backside of the correcting lens is a small circular convex mirror. As with a classic cassegrain telescope, the light gathered by the primary mirror is reflected back down the tube again through an opening in the primary to come to a focus behind it. This combination of mirrors and lens in an SCT allows a relatively long focal system to be housed in a relatively short optical tube. The typical

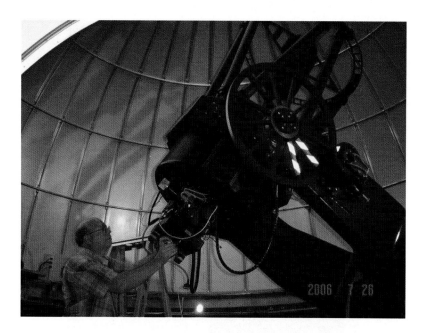

Fig. 7.3. The author behind a 24-in. (0.61 m) classic cassegrain reflecting telescope, a 2-ton professional instrument in a professional observatory owned by an amateur organization. (Credit: Trudy LeDoux/the Central Texas Astronomical Society).

8-in. f/10 SCT (Fig. 7.4) with a focal length of 80-in. can be contained in an optical tube only 22-in. long! This explains the popularity of the SCT; it is compact, relatively light, and easily transported! Several of my professional astronomer friends use these portable instruments on field expeditions. With an effective f/10 focal length, the SCT is a very good general-purpose telescope. However, the SCT is the least desirable telescope design for planetary observing. The compound lens design of the SCT requires a large secondary mirror. In an 8-in. f/10 SCT, this secondary mirror is 2.75-in. in diameter, causing a 35% obstruction of the primary mirror. This large secondary mirror causes a significant reduction in image contrast compared to refractors and Newtonian reflectors. For this reason, the SCT is not as good for visual observing. However, the SCT seems to perform as well as other telescope designs when used with CCD cameras or webcams for planetary imaging. Another problem with SCTs is image shift. Unlike other telescope designs, most commercially manufactured SCTs achieve focus by moving the primary mirror forward and backward in the optical tube. The mirror actually slides on an inner tube and is adjusted by turning a focus knob with a jack-screw that engages the primary mirror assembly. There must be a certain amount of play in this mounting so that the mirror can move without binding. If the amount of play is excessive the observer will notice a sideways shift of the image as the focus is changed. This image shift, while often slightly present in other telescope designs due to tolerances in the eyepiece rack and pinion focusing mechanism, can be very annoying in an SCT that is poorly assembled. I have experienced such large image shift in some SCTs that the primary mirror, when viewing near the horizon, has shifted enough off center as to no longer be

Fig. 7.4. An 8-in. (203 mm) Schmidt-cassegrain reflecting telescope on a fork-equatorial mount and tripod. (Credit: John W. McAnally).

adequately collimated! It would be very gratifying if commercial manufacturers would take pride in their product and truly eliminate this deficiency. In spite off its shortcomings, the SCT will remain popular. In its defense, I have made a high number of serious observations of Jupiter using an SCT. With care and patience, an observer using a Schmidt-cassegrain can produce fine results.

There are other types of telescopes, most of which are variations on the types already discussed. These other types include Maksutov, off-axis reflectors, Maksutov-Newtonians, Schmidt-Newtonians, binocular telescopes and others.

To summarize, the best telescope is one that produces a sharp image with high image contrast. This would favor refractors and long focal length (f/8 or longer) Newtonian reflectors. Size matters, and high resolution and high magnifications require large aperture. Therefore, a 4-in. refractor or 6-in. reflector is the minimum size for serious work. The optics of any telescope should be accurately aligned and collimated, since subtle features can be missed with a poorly working optical system. Finally, never put off an observing program for lack of the "perfect" telescope. Make the best observation you can with the instrument you have. In reality, the perfect telescope is one the observer uses well.

7.2 Eyepieces

The performance of the most modest telescope can be improved by using a really good eyepiece. Good eyepieces are important to planetary observing, since the observer is often using high magnification trying to see subtle surface markings. Thanks to modern manufacturing, there is a wide selection of very fine eyepieces to choose from today with a broad range of prices (Fig. 7.5). The eyepiece is one piece of equipment that the observer should not be too thrifty about. You should invest as much as you can in a couple of really good eyepieces. They can make all the difference.

To observe Jupiter, you should have a modest number of eyepieces to produce a range of magnifications. A good Barlow lens should also be included. Beginners often attempt to use too much magnification when observing Jupiter, thinking bigger is better. Experienced observers find this not to be true. Although, Jupiter is large and bright, it does not tolerate high magnification well – the image tends to go soft quickly. A good rule of thumb is not to exceed magnification of 40× per inch of aperture. With poor seeing, even that limit may not be attainable. With 8-in. telescopes, I rarely use more than 200× when observing Jupiter, even on nights of steady seeing. I have had observing sessions in which I achieved 325× usable magnification with my 8 in., but those nights have been rare. With Jupiter, try to go light on the magnification. You will learn from experience what you can achieve with your own setup.

As with telescope selection, arguments can arise over the choice of eyepiece design. An eyepiece used for observing the planets should be one capable of

Fig. 7.5. An assortment of eyepieces, both orthoscopic and plössl, and a ×2 barlow lens. (Credit: John W. McAnally).

producing tack sharp images. This normally dictates a multiple-element design; that is one that uses several lenses mounted in combination inside the eyepiece barrel. Planets are bright objects. Therefore, the eyepiece should have proper coatings to eliminate stray reflections and ghosting. Modern coatings actually increase light transmission through the eyepiece and contribute toward high image contrast. The field end of the eyepiece barrel should be threaded to receive eyepiece filters, to be discussed later.

My all time favorite type of eyepiece for planetary observing is the orthoscopic design. This eyepiece is a 4-element design. Orthoscopic eyepieces perform well at very high magnification, and produce a sharp image over almost the entire field of view. Their shortcoming is that they do not produce a very wide field of view, so they are not well suited for other types of observing. But, this is not a problem in planetary observing. Unfortunately, I do not see many offerings of orthoscopic eyepieces on the market today. A good used one can be a prized possession.

Perhaps the next favored eyepiece design is the plössl eyepiece. Certainly today this seems to be the most popular design, manufactured by many companies. The plössl eyepiece also uses a 4-element lens, but combines the lenses differently than an orthoscopic lens. Being very popular, plössl eyepieces are plentiful and come in a wide range of prices. These eyepieces have a fairly wide field of view and produce sharp images to the edge of the field, both at low power and high power. Plössl eyepieces are very good general-purpose eyepieces and also function well for close double star observing. Most of the eyepieces in my eyepiece box are plössl eyepieces. Here too, you should spend as much as you can afford on a quality eyepiece with good coatings.

Another common eyepiece type is the Kellner. Kellner eyepieces use a 3-element lens design. While Kellners perform well at low power for wide field viewing, they suffer more at high power than the previously mentioned designs. I do not use Kellner eyepieces for planetary viewing or other high power uses.

There are also multi-element designs using six or more elements in the eyepiece. These are usually eyepieces producing ultra-wide fields of view. They are also very expensive and I am not convinced they are any more useful for planetary work than orthoscopic or plössl eyepieces. Generally, this expense can be avoided.

There are other less expensive designs, many of which are 2-element lenses. These are poor for planetary viewing and should be avoided.

Although I prefer orthoscopic and plössl eyepieces for planetary observing, the observer can make up his own mind by trying various designs, especially the multi-element ones if so desired. Whatever eyepiece is used, it should be kept clean and free of smudges, and the lens elements should be held securely in place inside the eyepiece barrel.

7.3 Filters

Planetary observing is one endeavor in which the use of colored filters of specific wavelengths can be quite useful. We have previously discussed the fact that certain features on Jupiter tend to display certain colors, such as the bluish-gray festoons of the southern edge of the North Equatorial Belt, the redness of the Great Red Spot, or the reddish-brown coloration of the equatorial belts themselves. Filters can help us see these features.

Contrast and color differences between features on Jupiter are very subtle and especially difficult to detect by the inexperienced observer. Color filters (Fig. 7.6) can increase this contrast and help with the accurate determination of a feature's color. Filters can also help steady an image, especially when the seeing is poor or when attempting to view at a low elevation, such as near the horizon. An image that is unsteady or "boiling" due to poor seeing can be impossible to observe. Since different wavelengths of light are refracted, or bent differently as they pass through our atmosphere, the use of a filter to restrict the wavelength passing through to our eyes can improve this situation.

I strongly advocate the use of filters in planetary observing. This is a personal choice and some observers believe that filters reduce too much the brightness of the image. However, I believe most serious observers today would agree with me.

So, how do filters do what they do? A colored filter of the proper density will block all frequencies of light except for the one for which it is made to pass through. Simply stated, a red filter will filter out all wavelengths of light except the red wavelength, passing through the red light. Blue filters block all but blue light, and so on. We can take advantage of the transmission properties of these filters. For example, when observing a planet, the effect of a red filter is to make red features appear bright and other wavelengths to appear darker. The wavelength of the filter used determines which colors will appear darker. In general, to increase the contrast of a feature you want to observe, use a filter opposite the color of the feature. For example, to increase the contrast of a blue feature, use a red filter. In this manner, filters also assist with the identification of features. For example, if a feature becomes darker when viewed through a red filter, then the feature probably trends toward blue wavelengths. And, if it appears brighter through a red filter, then the feature

Fig. 7.6. An assortment of color filters, each mounted in a ring made to screw into the base of an eyepiece, shown for comparison. (Credit: John W. McAnally).

is probably red in color. This is a much more objective way to confirm the color of a feature, as opposed to viewing in integrated light since different observers can perceive color differently. For example, if you find it difficult to discern where the preceding and following edges of the GRS are, try using a blue filter.

Red filters, such as Wratten 21 (light orange-red), 23 (light red), or 25 (red) can assist in the identification of blue features such as the projections, festoons, and the bluish-gray features on the southern edge of the North Equatorial Belt. Red filters can also help you see really subtle features, like the south temperate oval BA, especially when a collar of bluish-gray material surrounds the oval.

Blue filters, such as Wratten 82A (light blue), 80A (medium blue), or 38A (blue) can be used to enhance red features, such as the Great Red Spot and the reddish-brown equatorial belts themselves. Since blue filters enhance, or darken, the equatorial belts, they help increase the contrast of bright features imbedded in the belts such as bright ovals and rifts. This is especially helpful when trying to accurately identify and measure the preceding and following ends of a rift or when measuring the length of a large bright oval.

I like to use yellow filters such as Wratten 12 (medium yellow) and 8 (light yellow) to observe the polar regions. Many veils and shadings have been observed in yellow light. South temperate oval BA can often be enhanced using the Wratten 8 filter, to distinguish it against an otherwise grayish background. I especially like the Wratten 8 filter as a general-purpose contrast enhancer. For me, this filter seems to increase overall contrast gently without washing out the more subtle features. I have also used green filters effectively. Wratten 11 (yellow green), 56 (light green), and 58 (green) will also enhance red and brown features.

Filters are usually threaded so they can be screwed into the field end of the eyepiece barrel. As with eyepieces, filters should be of good quality, squarely and securely mounted in their mountings, and optically flat on both surfaces. Filters should be kept clean of lint, dirt, and smudges so as not to obstruct the fine detail you are attempting to observe.

Again, there are varied opinions about filters, and the you should experiment to see which filters work best for you. Depending upon observing conditions, observing without a filter can sometimes prove to be the best choice.

7.4 Mountings

Mountings for astronomical telescopes come in various designs. These include alt-azimuth mounts, dobsonian mounts, german-equatorial mounts, and fork equatorial mounts. Regardless of the type, the mounting for the telescope should be sturdy enough to stabilize the telescope and hold it rigidly for viewing. Nothing is more frustrating than a good telescope on a poor mount that allows the telescope to vibrate and wiggle at the lightest touch. With such a mounting the image will never appear still enough for serious observing and the effort may as well be abandoned. Be wary of cheap, inadequate mountings.

I have observed with many different types of telescopes mounted on various mountings. Serious work can be performed with any type as long as it supports the telescope adequately. However, I strongly prefer equatorially mounted telescopes that are motor driven or other types that can be programmed to follow Earth's rotation. While an alt-azimuth, dobsonian, or hand driven equatorial mount can be moved by hand to place the object in the field of view, this is less than satisfactory. The continuous interaction with the telescope to keep up with the planet, especially when using high magnification, is tiring and consumes valuable observing time and hinders concentration. Equatorial mountings allow the telescope to rotate with the rotation of the earth and keep an object in the telescope field of view while moving around only one axis. The declination axis allows the telescope to move in the north–south direction, and the polar axis allows the telescope to move in the east–west direction. The polar axis is aligned with Earth's polar axis. The two main types of equatorial mountings are German-equatorial mountings (Fig. 7.7) and fork-equatorial mountings (Fig. 7.8). I also prefer mountings that are equipped with electronic slow motion controls so the planet can be easily centered in the field of view without touching the telescope by hand. Sometimes it is desirable to sweep the planet back and forth ever so slightly while viewing to help the eye detect subtle features. This sweeping is easily performed with no vibration transmitted to the telescope when using slow-motion controls. This technique is also effective for finding what I refer to as the "sweet spot" in the field of view. This varies with eyepiece and eye from observer to observer.

Mountings that are overbuilt are preferred over ones that are much too small. This may seem like an obvious statement, but heavier mounts perform much better

Fig. 7.7. A strong German-equatorial mount on a portable pier, made to carry a heavy telescope. The counter weight shown weighs over 50 lbs. (Credit: Dave Eisfeldt/the Central Texas Astronomical Society).

Fig. 7.8. A close up of a fork-equatorial mount, complete with setting circles on both the polar and declination axis. Electric clock-drive motors are concealed in the large circular base. (Credit: John W. McAnally).

in wind and breezes and dampen out vibrations more quickly. Generally, a mount that takes more than 3 s to dampen out vibrations and settle down is inadequate. Anti-vibration pads placed under the feet of a portable mounting can help reduce vibrations. You should pay attention to maintenance and keep all nuts and bolts in the mounting properly tightened. Adding mass to the mounting by hanging a can of sand or a large container of water from the mounting can also be helpful.

I also like to perform a fairly accurate polar alignment when using equatorially mounted telescopes. A proper alignment reduces declination drift, which will cause the object to drift north or south slightly over time, again requiring continuous adjustment. Observing sessions will be more enjoyable if these problems are avoided.

The telescope mounting is just as important as the telescope itself. When purchasing a telescope, you should test the mounting just as seriously as you would the optics of the telescope!

Chapter 8

Sky Conditions

8.1 Seeing

As astronomers we are always concerned with "seeing". The term "seeing" refers to how still the air is. When we look at a planet or other astronomical object, we look up through layers of Earth's atmosphere. This atmosphere is always moving. Light from the planet or star enters the atmosphere on its way to our eyes and is affected by these layers of atmosphere, which bends the light slightly in different directions as it travels toward us. This bending of light causes starlight to "twinkle" and the image of a planet to dance around and "boil" in our eyepieces. Have you ever noticed on a clear winter night after a relatively warm day, how the stars twinkle so vigorously? If you look at a bright star through a telescope on a night like that, you will notice how the star is not still at all. In fact, it may seem to bubble and roil, giving off sparkles of different colors. That is an example of poor seeing. If you were observing a pair of very close double stars, you might not be able to separate the two because they are so stirred up! Such poor seeing also affects planetary observing. Under the same conditions, the image of Jupiter in your eyepiece may appear to 'boil' and will not hold still no matter what you do. Of course, what is happening is radiative cooling. During the warm day the Earth absorbs heat. Then, when the Sun sets, this heat begins to radiate into the air as we cool down. These rising air currents also contribute to unsteadiness in the atmosphere.

Dr. Julius Benton gives a more formal definition of 'seeing'. According to Benton, "Astronomical seeing is the result of a number of very slight differences in the refractive index of air from one point to another, and such variations are directly related to density differences, normally associated with temperature gradients from one location to another. The observed effect of such random atmospheric deviations is an irregular distortion and motion of the image [511]." It is difficult to make a serious observation of Jupiter that will produce good data or images when seeing is poor.

Organizations like the A.L.P.O., B.A.A., and others have established arbitrary scales to estimate the seeing conditions during an observation. The A.L.P.O. uses a scale of 0–10 to indicate seeing conditions. Basically, the scale runs from 0, which is the worst possible seeing, to 10 which is perfect. Few of us ever see perfect seeing conditions. The A.L.P.O. scale gives some guidelines for these different levels of seeing and when useful work can be performed:

1. 0; the worst possible seeing, the image is completely unsteady and no fine detail can be made out.
2. 2–4; the whole disk moves with little detail made out.
3. 5; the planet's disk is stationary but boils, as though viewing the image through a moving liquid (Benton). Some fine detail can be intermittently seen.
4. 6–9; the planetary disk is stationary with little scintillation and fine detail can be made out.
5. 10; perfect seeing. The image is perfectly still. Smaller features normally not seen are easily made out.

Normally, a serious observation under high magnification should not be attempted unless the seeing is at least "5" or better.

Although poor seeing may be difficult to overcome, there are steps we can take to make sure we avoid conditions that would make seeing worse. We have just discussed that as heat rises, currents are put into motion that cause seeing to deteriorate. We want to avoid observing from locations that contribute to this problem. Keep in mind that stone, concrete, brick walls, and patios are all structures that act like a heat sink, storing up heat during a sunny day and radiating it away long into the night. Therefore, you should avoid setting up a telescope on a brick or concrete patio, or next to a building or street. Certainly do not set up a telescope on a concrete driveway or sidewalk, and try to avoid observing over rooftops or fireplace chimneys. As you see, with a little thought, some of this becomes common sense. If observing from inside a shelter, such as an observatory dome or roll-off roof facility, the temperature inside the shelter should not be different from the temperature outside. If observing from a dome in which the air inside the dome is warm from the day's heating, air will rise through the dome opening for hours until equalized. Have you ever gazed over a barrel of burning leaves at something in the distance? Remember how wavy everything was in your field of view. The heated dome will produce the same effect. Shelters should always be opened early enough prior to observing so the temperature inside can equalize with the outside air (Fig. 8.1). Large fans to move air during the day may be needed to help with this problem. Even outdoors, we often find that seeing improves as the night wears on and things cool down. I have often found the best seeing for observing Jupiter occurs after 2:00 A.M.

And so, we are not only concerned with the observing environment up high in Earth's atmosphere, but also the immediate environment around our telescope. Even telescope design can play an important part in "seeing". A refractor or a closed tube reflector like a Schmidt-cassegrain that has been stored indoors during the day will take considerably more time to cool down than a Newtonian reflector whose optical tube is an open truss system. With the closed tube, wavy air currents inside the tube can result in poor seeing until the tube cools down. Poor seeing not only affects visual observations, but photography and imaging as well, which we will discuss later.

8.2 Transparency

The other major atmospheric condition that astronomers worry about is "transparency". "Transparency" is a very descriptive term that simply refers to how clear the sky is. Most observers measure transparency by simply using

Fig. 8.1. The professional observatory of the Central Texas Astronomical Society. Notice the hood on the side that conceals a large exhaust fan. One on each side of the building helps to minimize the buildup of heat inside the dome during the day. (Credit: John W. McAnally)

the stellar magnitude scale, and this is a system of measurement advocated by the A.L.P.O.

A measure of transparency simply states the limiting magnitude that can be seen in the immediate vicinity of the object being observed, absent light pollution. In other words, transparency indicates clear sky or clouds in varying degrees. For example, if during an observation under dark skies sixth magnitude stars can be seen, then we say the transparency was six. If only third magnitude stars could be seen in the vicinity of the object being observed, we would say the transparency was three, and so on. When observing from a light polluted area, you will need to make a judgement as to what the transparency would be if you were under a dark sky. And so again, the use and application of the transparency scale is common sense.

Along with transparency, other conditions that have an affect on our observations would be the presence of haze, water vapor in the air, or wind. Transparency could be quite good under each of these conditions, but our ability to make a good observation could still be degraded. Recording this type of information is important, as we will see later in this book.

8.3 Learning about Weather

How many times have you experienced a wonderful afternoon and, thinking the night would be good for observing, set up your equipment only to discover the seeing conditions were horrible? Or, having had a clear day, find clouds rolling in just minutes after darkness? I have experienced this disappointment more times than I care to remember. Too often it seems to happen when I have planned a very important, once in a lifetime, observation. What can we do about this dastardly state of affairs? Sadly, often is the case we can do nothing at all. But sometimes we could have averted disappointment if we simply understood better our own weather. More specifically, I am referring to cause and effect in weather.

Now, it is not my intention to turn you into an expert weather forecaster; we'll leave that to the professional weather people to attempt. However there are some phenomena each of us can better understand and make use of.

I have already mentioned the cooling that occurs at night after a warm day. This problem is especially pronounced when the day has been cloudy, followed by a very clear night with no cloud cover. We can almost predict without fail that, for at least several hours into the night, the seeing will be horrible in spite of the perfect transparency. An opposite condition can also occur. Sometimes, after a front has moved through the area following a rainfall clearing the sky, the seeing conditions can be superb.

Many of my colleagues and I have discovered that a slight haze in the sky can actually result is very good seeing in which the planet's disk is surprisingly still, resulting in an effect opposite that of radiative cooling. Of course, some of the finer planetary detail can be obscured if the haze is too thick. Regardless, try observing under these conditions the next time you have a chance. One way to do this is to observe when there are intermittent clouds present. The planet's disk will periodically disappear into this haze, then cloud, and back into view again. You will be surprised to find that what you thought was a useless night produced some of your better observations. You can study books about weather and pay attention to how different weather conditions affect the seeing in your area. Keep a log of your local conditions and try to remember when a weather change has resulted in good observing conditions.

Chapter 9

Making a Record

At some point, if you are like most amateur astronomers, you will want to record what you are seeing through your telescope. The desire to record all the wonderful things you can see is almost irresistible, but how? Many of you will first try to draw what you see. Then you'll wonder if your simple film camera can do the trick. Finally, after some searching around, you'll discover the more sophisticated imaging devices available today. So, how do you get started? Well, that is our next discussion.

9.1 Making a Drawing of Jupiter

If you wish to make a serious study of Jupiter you should observe it as often as possible. Only when you stay abreast of conditions on the planet can you hope to recognize a subtle change when it occurs. If you observe only occasionally you will never know whether something you see has any significance. Nor will you know when an event began or how long it has run.

Drawing or sketching what you see while looking through a telescope is the oldest and simplest method of making a record of the observation. Even today, in this high tech world, the eye as detector and paper and pencil as recorder is a simple, inexpensive valuable method available to everyone. Not necessarily to be outdone by CCD cameras, modern observers with very high acuity of vision like Carlos Hernandez, John Rogers, Claus Benninghoven, and Stephen James O'Meara can accurately record amazing detail with just their eyes and a telescope of good quality. Indeed, during the era of photographic film, the eye was always able to see more planetary detail than film could capture. Studying Jupiter by making a drawing has another attribute. The process of drawing the planet will make you a better observer; it trains the eye to see fine detail. It trains the mind to remember what has been seen. Somehow, the simple act of drawing what we see puts us on a higher intellectual plane. Even if you feel that you are not skilled at drawing, you should give it a try. In fact, there are techniques that can help even the beginner. And remember, not only is it pleasurable to record what we see, we are also trying to make a record that has scientific value.

The drawing of Jupiter should present the planet in the normal configuration as seen through an astronomical telescope, inverted with the south pole at the top. This is easily done in Newtonian reflectors. However, refractors and Schmidt-cassegrains often come supplied with star diagonals that make it easier for the observer to view objects near the zenith. Use of these star diagonals will insert an extra reflection, an odd number of reflections, into the light path of the telescope.

The effect of this is to give a view in which south is down and east and west are reversed. You should avoid using star diagonals and make the drawing the proper way, so that south is at the top. If you insist on using a star diagonal, then the observing form should prominently note this use and indicate north, south, east, and west on the drawing as seen through the star diagonal.

We must remember that a drawing is not a photograph or digital image. It is therefore subjective, being what we perceive with our eyes and our mind, and what we record with our hand. Consequently, as observers we must make an honest attempt to record only what is truly seen, and draw only the amount of detail we can see. We must never record what we think we see or, more importantly, what we want to see or think should have been seen. A dishonest observation will be found out and the integrity of the observer will be destroyed. In science, the loss of integrity is very difficult to restore. Sometimes, it is acceptable to record a feature that is "suspected", and this condition should be prominently noted in the data. In spite of the inability of film to capture fine detail, one of its advantages was that it was an objective, unbiased means of capturing the appearance of the planet. We will discuss the use of film later.

There are two basic types of drawings, a full disk drawing and a strip sketch.

9.2 The Full Disk Drawing

In making a full disk drawing, the process should be completed in a series of logical steps. First, use a preprinted observing form that has Jupiter's oblate outline already recorded (Fig. 9.1). There is no scientific value in practicing to draw the perfect disk, and your observing time can be better spent drawing details anyway. The B.A.A and A.L.P.O. have observing forms that anyone can obtain and use. Forms can be reproduced onto heavier drawing paper that will hold up better under nighttime dewing and the inevitable erasures. These forms also provide a place to record data about the observation such as time, weather, seeing and transparency, telescope, eyepieces, filters used, and observing location. All of this information is important to give credibility to the observation. A selection of good artist pencils in various grades of lead are also helpful, as will be a good paper stump for smearing, and a good artist's or architect's eraser for erasing cleanly.

To begin the drawing, you should first observe the planet for several minutes, noting the features that are present. Be patient, use different magnifications and different filters, and see how the planet is going to perform for you. By all means, get into a comfortable viewing position (Fig. 9.2). Prepare yourself for an adventure!

Because Jupiter rotates so rapidly upon its axis, it is obvious in just 20 min that features are moving across the disk as the planet rotates! Therefore, it is imperative that a drawing of Jupiter be completed in 20 min or less. This task seems daunting, but it can be done. If a drawing takes more than 20 min, the placement of features can be skewed and inaccurate.

With form in hand begin the drawing by first sketching in the positions of the north and south equatorial belts. Accurately placing these belts on the drawing is critical and can be a little tricky. Poor placement can adversely affect the subsequent placement of other features during the remainder of the drawing. Bertrand Peek commented on this critical placement of the belts and wrote, "...for it is essential that in making a drawing the conspicuous markings should be used as fiducial points, to which the fainter detail can be related. The author (Peek) confesses that

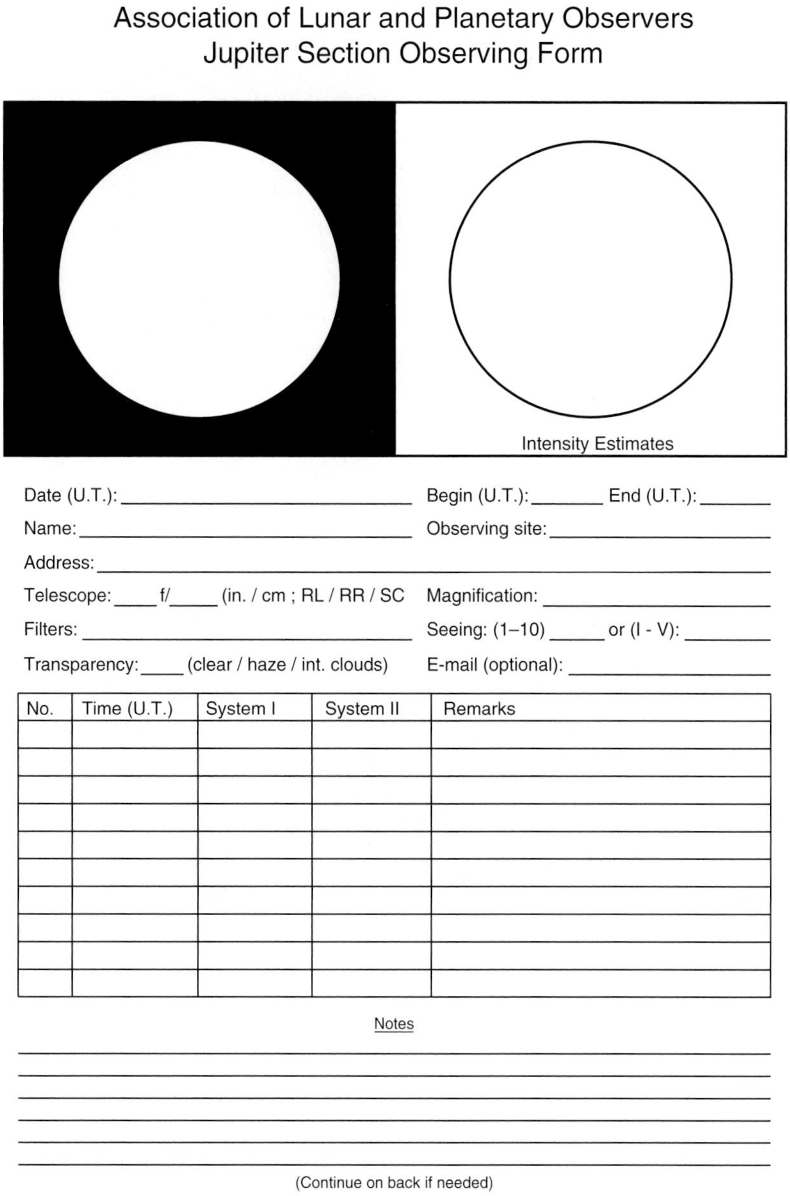

Association of Lunar and Planetary Observers
Jupiter Section Observing Form

Intensity Estimates

Date (U.T.): _____   Begin (U.T.): _____ End (U.T.): _____

Name: _____   Observing site: _____

Address: _____

Telescope: _____ f/_____ (in. / cm ; RL / RR / SC   Magnification: _____

Filters: _____   Seeing: (1–10) _____ or (I - V): _____

Transparency: _____ (clear / haze / int. clouds)   E-mail (optional): _____

No.	Time (U.T.)	System I	System II	Remarks

Notes

_____
_____
_____
_____
_____
_____

(Continue on back if needed)

Fig. 9.1. An observing form used to make a full disk drawing of Jupiter and record intensity estimates and the transit times of features as they rotate past the central meridian. (Credit: the Association of Lunar and Planetary Observers).

he has always hesitated to embark upon the making of a drawing of Jupiter, unless there has been to hand a fairly recent series of latitude measurements that will ensure his placing at least the four main belts at there correct distances from the centre of the disk. Some draughtsmen have a personal tendency to place objects too far from the centre of a planet, others to concentrate them there; so it is wise to resort to any legitimate means of attaining reasonable accuracy [512]."

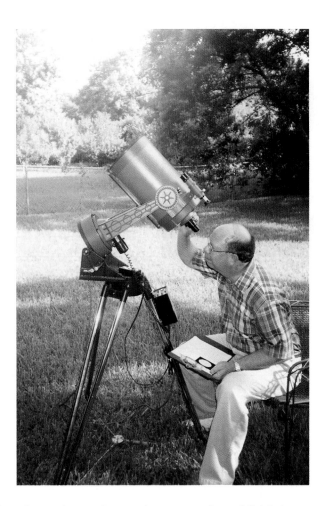

Fig. 9.2. The author at telescope during early evening, making a full disk drawing and recording transit timings. (Credit: John W. McAnally).

I have often used an eyepiece with a graduated reticle pattern to measure the disk of Jupiter and note the placement of the belts in relation to the reticle pattern. To do this, first measure the planet from pole to pole, noting the number of graduations over that distance. Transfer this number of graduations onto one of the observing forms. Next, lightly sketch in the positions of the belts in relation to that reticle pattern. Finally, go back to the eyepiece with the observing form and make sure you are satisfied with your placement of the belts. This form can now serve as a master form to lightly sketch in the belts on future observing forms. Usually, these measurements are good for several months. The measurement of the planet with the reticle pattern eyepiece can be performed several times during the apparition. Once the belts have been accurately placed on the form, other large features should be sketched in relation to each other. The shadings of these features, dark and light, can be partially completed.

Next, smaller features that can be seen should be drawn in relation to the positions of the larger features. By this time, you should be nearing the end of the 20-min period. During the time spent so far, you may have viewed through red, green, and blue filters.

Finally, more detail of all features already drawn can be added, including an accurate representation of the darkness or brightness of the feature.

In the remaining couple of minutes, compare the drawing to the telescope view making sure the drawing is accurate. Make note of the ending time of the drawing in universal time. Recording this time is very important.

Now, you can relax and record the other data needed for the observation (Fig. 9.3). Every drawing must record some standard information about the observation you made. For example, the A.L.P.O. observing form records the following information:

1. The name of the observer.
2. The location the observation was made from and the address (contact information) of the observer.
3. The date and time of the observation in Universal Time. Because your data may be of interest to astronomers all over the world, it is important to record this date and time in a manner that will not be confusing since people from different countries record dates in different formats. So, I always spell out the month, etc., like this: 2004August05/12:15 U.T. Remember that the date must also conform to universal time.
4. The type of telescope and its aperture.
5. The magnifications used during the observation.
6. Filters used, if any.
7. The seeing and transparency.
8. The beginning and end time of the observation in universal time.
9. The longitude of the planet's central meridian at the end of the observation/drawing.
10. And space to record other information about observing conditions such as the presence of haze, wind, intermittent clouds, and any other descriptive information the observer feels may be necessary to help the reader fully understand the observation/drawing.

To the beginner this probably seems like an enormous amount of data to record for a drawing that was completed in just 20 min. But the truth is, without this information the drawing you just made is almost worthless. Next to the accuracy of your drawing, this information is the most important part of the observation. It allows the observation to become part of the historical record of Jupiter, and it allows other scientists to evaluate the data for themselves, a very important part of the scientific method. It also allows others to form an opinion about the credibility of the observation. An observation made under favorable conditions with adequate equipment is certainly more believable than one made under impossible conditions with a telescope that is too small to reveal the features recorded. So by all means, record the required information for your observation so that others will be able to rely upon your work.

As we will see, regardless of the type of observation or imaging you perform this information will be required of you.

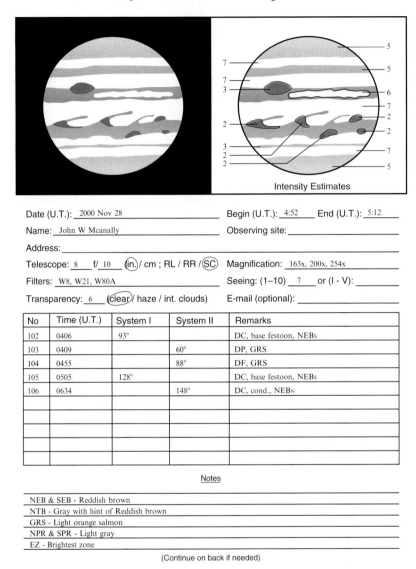

Association of Lunar and Planetary Observers
Jupiter Section Observing Form

Intensity Estimates

Date (U.T.): _2000 Nov 28_____ Begin (U.T.): _4:52___ End (U.T.): _5:12____

Name: _John W Mcanally_____ Observing site: _____

Address: _____

Telescope: _8___ f/ _10___ (in.)/ cm ; RL / RR /(SC) Magnification: _163x, 200x, 254x_____

Filters: _W8, W21, W80A_____ Seeing: (1–10) _7___ or (I - V): _____

Transparency: _6___ (clear)/ haze / int. clouds) E-mail (optional): _____

No	Time (U.T.)	System I	System II	Remarks
102	0406	93°		DC, base festoon, NEBs
103	0409		60°	DP, GRS
104	0455		88°	DF, GRS
105	0505	128°		DC, base festoon, NEBs
106	0634		148°	DC, cond., NEBN

Notes

NEB & SEB - Reddish brown
NTB - Gray with hint of Reddish brown
GRS - Light orange salmon
NPR & SPR - Light gray
EZ - Brightest zone

(Continue on back if needed)

Fig. 9.3. A completed full disk drawing with related information. (Credit: John W. McAnally).

9.3 The Strip Sketch

The strip sketch shares many of the same characteristics of the full disk drawing, but differs in one important respect (Fig. 9.4). In making a disk drawing, the observer concentrates on a relatively narrow range of latitude, and does not draw the full disk. As such, a strip sketch allows you to concentrate in more detail on a particular

Association of Lunar and Planetary Observers
Jupiter Section Strip Map

Date (U.T.): _____ Begin (U.T.): _____ End (U.T.): _____

Name: _____ Observing site: _____

Address: _____

Telescope: ____ f/____ (in. / cm. ; RL / RR / SC   Magnification: _____

Filters: _____ Seeing: (1–10) _____ or (I - V): _____

Transparency: ____ (clear / haze / int. clouds)   E-mail (optional): _____

No.	Time (U.T.)	System (I)	System (II)	Remarks

Fig. 9.4. An observing form for making a strip sketch of Jupiter and for recording transit timings. (Credit: the Association of Lunar and Planetary Observers).

part of Jupiter's disk, since you are not devoting valuable observing and drawing time to the whole planet.

For example, a strip sketch might be done that only records a section of Jupiter from the latitude just south of the South Equatorial Belt to just north of the North

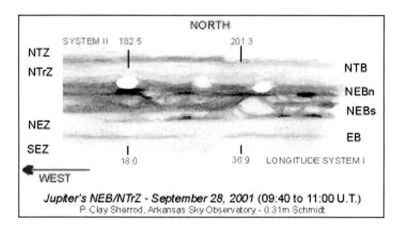

Fig. 9.5. A wonderful strip sketch of a portion of Jupiter made by Clay Sherrod. Note the longitudinal data provided and the labeling of the belts and zones. (Credit: P. Clay Sherrod).

Equatorial Belt. It can span the entire width of the planet or a shorter segment. If desired, the strip sketch can be even more restricted than this, covering only the Great Red Spot (GRS) and its immediate surroundings, for example, allowing you to concentrate on this in great detail. Phillip Budine refers to these as sectional sketches, covering only a short longitudinal section of the planet. Or, the strip sketch can be kept going for hours, making a continuous record of the planet as it rotates during the night. When this latter method is adopted, the sketch can be done in conjunction with central meridian transit timings, to be discussed later, and tied to the sketch by annotating features with their longitude. In recent times, Claus Benninghoven and Clay Sherrod have demonstrated great skill in making strip sketches, rivaling information recorded by CCD cameras (Figs. 9.5–9.7).

To start a strip sketch, all the rules discussed in making a full disk drawing should be followed. Limit the observation to no more than 20 min for the section being drawn, recording the large features first, then drawing in the smaller features in relation to the larger ones. Finally, check the accuracy of your sketch with the view in your telescope. The A.L.P.O. also provides a form for the strip sketch. The same data mentioned above for a full disk drawing should also be recorded for a strip sketch.

9.4   Intensity Estimates

When completing a drawing of Jupiter, observers can also makes estimates of the intensity of the features recorded. While the value of a drawing is its ability to provide a visual record of the planet's appearance, intensity estimates provide a source of data that can actually be quantified. This is very important data that the amateur can collect. Professionals find this data, collected over time, to be valuable to their own research.

Observing forms can also provide a place to record intensity estimates (Fig. 9.1). The observing form used by the A.L.P.O. provides a separate disk on its form next

HISTORY OF NEB DISTURBANCE
JUPITER - Sept. 23 to Oct. 5, 2001
P. Clay Sherrod, Arkansas Sky Observatory

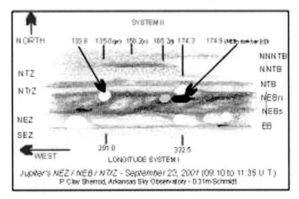

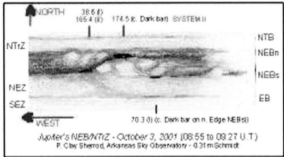

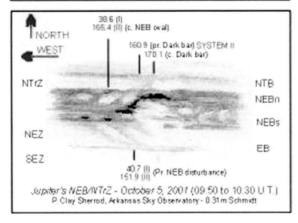

Fig. 9.6. More amazing strip sketches of Jupiter by Clay Sherrod. Note the spectacular detail he has recorded! (Credit: P. Clay Sherrod).

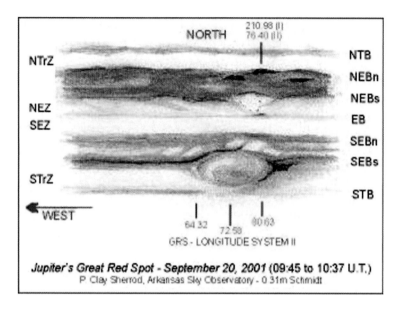

Fig. 9.7. Jupiter's Great Red Spot and surrounding area, captured in a strip sketch by Clay Sherrod. Note that Clay made transit timings of the preceding edge, center, and following edge of the GRS. (Credit: P. Clay Sherrod).

to the disk provided for drawing the planet. Intensity generally refers to the relative darkness or brightness of a feature. The A.L.P.O. uses a numerical scale to indicate the relative intensity of features seen.

The A.L.P.O. intensity scale is a numbered scale from 0 to 10, with 0 indicating the darkest feature and 10 the brightest. Many years ago, Phillip Budine stated these intensity values as:

10 Unusually brilliant zone
 9 Extremely brilliant zone
 8 Very bright zone
 7 Bright zone
 6 Slightly shaded zone
 5 Dull zone
 4 Dusky polar belt
 3 Dark belt
 2 Very dark belt
 1 Extremely dark belt
 0 Black, shadow of a planet

With the use of this scale, you can estimate the relative brightness of belts and zones and individual features, such as condensations, festoons, and bright ovals (Fig. 9.3).

With experience, you can estimate the intensity of features with good accuracy. The ability to make intensity estimates certainly improves with practice but might appear intimidating to the beginner, who might have difficulty getting started. In my own experience I have found, at least during recent apparitions, the general overall intensity of the North Equatorial Belt (NEB) to be 3, with the relative brightness of the North Tropical Zone to be 7. Condensations along the north edge of the NEB to be 2, and bright ovals imbedded in the NEB to be 7.5–8, and so on. You should not approach these intensity estimates with any preconceived notion, but my description here can provide an idea of the process.

The application of the scale is certainly subjective, and one observer might estimate the intensity of a feature to be slightly brighter than another observer. However, with practice, you can learn to be consistent in the application of the scale as you become familiar with the appearance of the planet. In spite of this subjectivity, the intensity scale is a much more quantitative and accurate method than simple word descriptions, which can be rather vague or ambiguous at best. This quantitative scale has another major advantage; it can be statistically analyzed over time. And, in spite of the subjectivity, the effect of personal bias can be minimized when intensity estimates by a large number of observers during the same apparition are included in the statistical analysis.

9.5   Central Meridian Transit Timings

Of all the visual observing programs, the central meridian transit timing is the most valuable observation the amateur can make. The data produced by central meridian (CM) transit timings is extremely important to the professional community because, above all other amateur endeavors, it contributes to the understanding of Jovian wind currents, jet streams, and weather. Here, the amateur truly makes a meaningful contribution.

When properly performed, CM transit timings result in data that provides the longitudinal position of features in Jupiter's cloud tops. By recording the positions of features over time, the speed of wind currents and jet streams at various latitudes can be determined. All that is needed is a steady telescope, an accurate timepiece, patience, and dedication. The transit timing is easily accomplished but requires an understanding of some basic principles.

The Central Meridian is an imaginary line that runs from Jupiter's north pole to its south pole, evenly dividing the planet (Fig. 2.1). Like the Earth, Jupiter is divided into 360° of east–west longitude. By referring to an ephemeris, it is possible to determine, for any given minute, the longitude that happens to be on Jupiter's central meridian at a given time. For any feature observed on the CM, if we determine the longitude of the CM, we also determine the longitude of the feature itself at that date and time. Simple enough!

Some advanced amateurs may use a filar micrometer that is available to them to mark the central meridian. However this expensive piece of equipment is really not necessary. For years I have made this measurement with nothing more than my eyes. I simply make a visual estimation of the CM noting the crossing of features over my imaginary line. With practice, it is possible to become quite good at this, and it is always satisfying to see your timings in agreement with other observers around the world!

Of course, noting the crossing of a feature across the CM is also a bit subjective. An observer might be uncertain and delay the marking of the transit for several minutes, in error. What then may be done to minimize this problem? Phillip Budine, past A.L.P.O. Jupiter Section Coordinator, recommends what I refer to as "the three minute rule." Simply put, the human eye may perceive an object to be on the CM for up to 3 min. In making your transit timing, if you note the minute at which the feature first appears to be on the CM, and then note the time that the feature is obviously no longer on the CM, then an average of these two times will usually prove to be the best estimate of the transit. I have used this "three minute rule" extensively and it has always produced reliable results. The goal is to note the transit to the nearest minute. Therefore, a timepiece or some other means of noting time to the nearest 30 s, such as the time signals on a short-wave radio, is required. When using my watch, I always set it first to the time signals on the short-wave.

Of course, the timing just made must be recorded, and the forms used for making sketches can also provide a place for this. The A.L.P.O. forms have a place to record this data. Once again, Universal Time should be used to record the time and date of the transit. The feature's longitude, either in System I or System II as appropriate, should also be recorded. As noted previously in this book, visually Jupiter has two systems of rotation, System I and System II.

A clear, straightforward description of the feature observed should also be recorded. Phillip Budine developed a system of nomenclature that is remarkable in its accuracy and simplicity. This system of nomenclature was discussed in Chap. 2 of this book. This method of notation, when properly used, will provide an abbreviated, yet accurate description of the feature, including its location among the belts and zones of the planet (Figs. 2.2 and 2.3). For example, the center of a dark condensation observed to transit the CM on the northern edge of the North Equatorial Belt would be described as: Dc, sm cond., n edge NEB. Similarly, the center of a bright oval observed in the center of the South Temperate Belt would be described as: Wc, oval, center STB. And therefore, the descriptions used would include D for dark, W for white, p for preceding end, c for center, and f for following end. Along with these notations would be recorded the System I or II longitude of the feature and the Universal Time of the transit. The System I or II longitude can be determined and entered onto the observing form at a later time, when the ephemeris is consulted. An ephemeris can be obtained from any number of sources today, including a web site maintained by the Jet Propulsion Laboratory. Also, there are computer programs available that can yield the CM longitude when the date and time in Universal Time is entered.

I have spent many a wonderful night behind a telescope watching Jupiter rotate, recording transit timings for hour upon hour. I find it very satisfying to collect data that I know is valuable to the study of Jupiter. And there is the sense of the hunt, waiting for the next feature to appear from around the following limb of the planet. There is always the anticipation of what may appear next!

9.6 Drift Charts

After recording a series of central meridian transit timings, it will be useful to plot these positions on a chart so that the drift rates of features can be determined. A drift chart also helps us see how features behave in relation to each other and it helps us predict their future behavior and position. Thus, the "drift chart" becomes a valuable part of the scientific record for an apparition. A drift chart is also referred to as a "drift line graph."

To complete a drift chart by hand, the longitudes of features are entered on the chart against the dates of observation. We normally find the chart to be most easily interpreted when the date is represented by the vertical axis and longitude is represented on the horizontal axis. Almost any graph paper purchased at an office supply store will suffice. The graph paper should have enough graduations so that it is relatively easy to plot longitude along the top margin of the paper to the nearest degree, even when interpolating between lines. Running vertically, down the left margin of the paper, a distance of about two inches per month normally provides enough room to accurately represent the days of the month. With a little practice, setting up the graph becomes an easy task. Since the apparition will run for several months, continuation pages will be necessary. You can also use software programs to make measurements of images. *Jupos* is a piece of software with amazing capabilities that is easily located by searching the Internet.

A typically plotted drift chart is illustrated (Fig. 9.8). Here we see that features drift in decreasing longitude with time, giving the plot the appearance of a slanted line. Sometimes features can drift at inconsistent rates or may pause altogether for a short period of time, or they may speed up. It is therefore desirable to have enough observations so we can have confidence in the chart being produced. A chart with too few observations may be suspect, especially if the results seem to indicate a varying drift rate. With too few observations you cannot be certain if the feature is really speeding up or slowing down, or if there is simply error in the CM timings themselves. A large number of observations will tend to cancel out random errors in timings. As illustrated, I like to plot the observations in the form

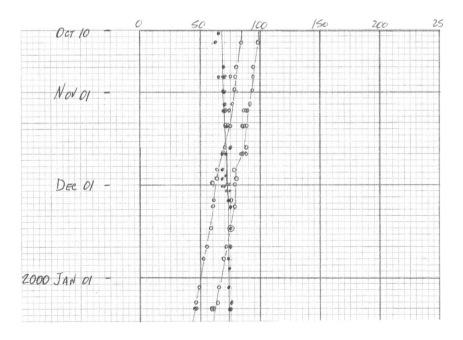

Fig. 9.8. A drift chart created by the author to reduce transit timing data. This one depicts a conjunction of the GRS and South Temperate Ovals BE and FA. Note how the drift rates of the ovals changed as they approached and then passed the GRS. (Credit: John W. McAnally).

of a scatter diagram, then using a least squares solution, construct a line through the plots to interpret the drift rate of the feature. I feel it is important to plot all of the timings since this adds to the credibility of the graph by allowing the reader to see the raw data that was collected and how it was interpreted. Presenting the data in this visual way is simple and straightforward, allowing anyone reading the graph to judge the deviation in the observations and to access the adequacy of the number of plots recorded. In the past, some observers have been criticized for not allowing all the plots to be seen, simply showing the interpreted drift line by itself, inadvertently precluding any scrutiny of the work. I believe it is more desirable and credible to plot all the observations, as shown. The illustration (Fig. 9.8) shows oval BA, a bright oval, overtaking and passing the GRS, a dark feature.

Since Jupiter is divided into System I and System II rotation rates, it is helpful to have several graphs to present like features together. For example, one chart might present the plots for features in the latitude from the middle of the South Equatorial Belt to the South –South Temperate Belt (SSTB) (System II), or for the region of the southern edge of the North Equatorial Belt to the northern edge of the South Equatorial Belt (System I), and so on. I normally subdivide the plot of features even more than this, to avoid overcrowding on the charts. One of the most fascinating behaviors to observe and measure is that of the GRS, south temperate oval BA, and the smaller ovals of the SSTB. In recent years, the GRS has been relatively stationary and plots of oval BA will show oval BA overtaking and passing the GRS. The drift rate of oval BA is affected by the GRS when it passes and this is easily seen when plotted on the drift chart. Likewise, the behavior of the SSTB ovals is also fascinating. But, more importantly, the data being collected is very important. I have previously related the story of the South Temperate Dark Spot of 1998. Transit timings plotted on a drift chart alerted us to what turned out to be a new, significant feature.

I also like to prepare a table displaying the dates of the transit timings, the longitudes, and the names of the observers to be published along with the drift chart. When research is published it is accepted procedure to disclose how the observations were made, the instruments used, and the data that was collected in arriving at the findings. Again, I believe publishing the data in this way helps provide complete disclosure to the astronomy community.

Once the transits have been plotted on the drift chart, it is possible to calculate the rotation rates of the feature located at that latitude, thus allowing the speeds of currents and jet streams to be determined for the apparition. This data, gathered from apparition to apparition, allows patterns of behavior to be determined. With this information, scientists can create computer models to analyze atmospheric conditions on Jupiter.

9.7 Observation and Estimates of Color

In Chap. 3 we discussed the observation of color in Jupiter's atmosphere and I encourage the reading of that section again. We discussed how subjective and inaccurate the observation of color can be and how some of our predecessors, pillars of the amateur astronomical community, have dealt with the subject. In spite of the difficulties, I must admit I consider the study of color in Jupiter's atmosphere to be one of the most pleasurable and intriguing parts of my passion for the planet.

The contrast of colors in Jupiter's atmosphere is very subtle. You may have particular difficulty discerning from among the different shades of gray, brown, alabaster, and ochre; and we have already discussed how the GRS isn't so red after all, at least at the present time! So, if the determination of color is so difficult and subjective, of what use is the endeavor?

I believe that, as with drawing the planet, the study of color trains the eye to see. It improves our skills and makes us a better observer. It will train the careful, honest observer to be all the more mindful of the integrity of his observation. Over time, the observation of color in a consistent manner can provide useful data regarding trends in Jupiter's atmosphere. We have referred to "coloration events" in Jupiter's Equatorial Zone. We have discussed the fadings and "Revivals" of the South Equatorial Belt and the subsequent increase and decrease in intensity of the GRS. Belts fade and are restored. All of these events are impossible to discuss without introducing some discussion of color. The great history of Jupiter is replete with the discussion of color by past observers. It is as though the discussion of color in Jupiter's atmosphere is ingrained in the human spirit and we should not deny it to ourselves.

It is my practice to record the color of Jupiter's belts, zones, and other individual features on the Jupiter observing form. I generally make this determination while I am making intensity estimates. I find that estimates of color, as with intensity estimates, are best performed at lower powers of magnification in integrated light; that is, without the use of color filters. With a Celestron C8 I usually do this at a magnification not exceeding 175×. With a 12-in. instrument I would use a higher power. Excessive magnification causes the image to be dimmer, and the color to be more diffused or diluted. The intensity of the color needs to be strong enough that your eye can make a reasonable estimate of its tint and intensity. I also like to record this information in the descriptive notes of my observing log, a log that I keep for all of my astronomy observations, including the non-planetary ones. As subjective as it is, we never know when our record of color will turn out to be the most important piece of information recorded on a given night.

9.8 The Use of Photography to Study Jupiter

Every serious observer of Jupiter will eventually be drawn to the fascination of photography and a desire to record the appearance of the planet in a more objective manner. When I was a boy, I imagined the day when I would possess that special 35 mm camera with which I would be able capture the stars and the planets on film. Since then I have done just that, and what I have learned has been somewhat frustrating. I know this must sound surprising and disappointing to those of you who have reached that same, magical stage in your astronomy life, but it is true.

Although film as a recording medium is completely objective, it does have its own shortcomings. It seems the "seeing" we discussed earlier plays havoc with film photography, especially when attempting to capture fine detail in small, planetary features. Older textbooks can be found that show photographs of Jupiter, Saturn, and Mars taken with the 5-m (200-in.) Hale Telescope at Mount Palomar on photographic plates that were available at the time. While these photographs reveal the larger, more obvious albedo features, the human eye was able to see much more detail just looking through an eyepiece in a smaller telescope. The Palomar photographs suffered from the effects of "seeing".

Unlike deep sky photography of stellar and nebulous objects in which the camera can be placed at the prime focus of the telescope, planetary photography requires some amount of "eyepiece projection", so that the image falling on the photographic film is large enough for the film to record planetary detail with satisfactory resolution. To achieve this projection, an eyepiece is placed in front of the camera, using a tele-extender between the camera and eyepiece so that the camera's film plane is some distance from the eyepiece, thus projecting an enlarged image on the film. Jupiter is a bright object and we normally do not have much difficulty observing it visually at moderately high magnifications. But when the image is enlarged to project onto film, the act of enlarging the image spreads out its light and also reduces its brightness. This dimming requires us to lengthen the exposure time to capture an acceptably bright image on the film. And there's the rub. When we lengthen the exposure time, we give the effects of poor seeing a chance to blur our photographic image. The exposure time required to capture an enlarged image of Jupiter can range from 1 to 3 s, and perhaps longer. During the 3 s or more of exposure, the waviness of the currents in the air smears the fine detail that we want to record. There are modern, fast films available that can reduce exposure time, but normally the faster the film the coarser the grain of the emulsion. This coarseness will reduce the film's ability to record fine detail.

Donald Parker, world-renowned amateur planetary imager, used film successfully into the late 1960s before the advent of CCD cameras. Don produced some of the best film based photographs of the planets taken up to that time and yet, he was not able to capture all that could be seen with the human eye.

Astrophotography with film has always been an exciting endeavor for me, and every amateur astronomer should give it a try. I have seen some surprisingly good film images taken of Jupiter by a few amateurs with near perfect seeing conditions. But be prepared for disappointment and a lot of darkroom work trying to improve contrast in the final print to squeeze out as much detail as possible. There are many good books on the subject and you should certainly have a few of them in your own library. Remember that the same information recorded when a drawing is made of Jupiter should also be recorded for the photograph, including information about the film, camera, and exposure.

Fortunately, technology marches on and advancements have led to a solution for the problems with planetary photography through the use of CCD cameras and web cams.

9.9 CCD Imaging

For the amateur astronomer, the development of the affordable CCD camera has been as important to the advancement of our science as big telescopes and high-speed photographic films were to astronomers of ages past. According to one professional astronomer whose specialty is instrumentation, the CCD camera is allowing amateurs to do scientific work that is as good or better than the work professionals were doing as recently as the 1970s.

There are numerous books available today on the subject of CCD imaging, written by real experts that discuss CCD imaging in great detail. So many books have been written by such smart people, that I would refer you to them and their knowledge of CCD imaging. One of the best books is one in the Series of Patrick Moore books, edited by David Ratledge titled *The Art and Science of CCD Astronomy*. Several talented CCD imagers contributed to the book and I highly recommend it. You can learn more from them than I can teach you in the few pages available here. However, it will serve us well to discuss the advantages of CCD imaging over film and the results that can be attained.

Like film, a CCD image is an absolutely objective record of the physical appearance of Jupiter and the placement of its features. The placement of its features; this is important! A CCD camera is so sensitive that it can capture an image in a fraction of the exposure time required for film, effectively freezing the image. Consequently, a CCD image suffers far less from the effects of "seeing" than film does. At best, film captures 3–5% of the light that lands on it. By contrast, a CCD chip captures 30–50% of the light that strikes it, a tremendous gain in light gathering ability [513].

The term "CCD" stands for charge coupled device. A CCD is a silicon chip. Originally developed as a storage device, it was discovered that a silicon chip is sensitive to light. And whereas a silicon chip is normally encased in a black plastic case to keep light out, a CCD chip is made with a window opening on its top to let light in. Silicon is sensitive to the visible and near infrared part of the spectrum. It is sensitive in the sense that it will convert incident light (photons) into an electric charge (electrons). The active light-exposure part of the CCD is divided into *photosites* or pixels in a matrix of rows and columns. Each photosite converts light (photons) into electrons and stores them until the end of the exposure. The number of photons produced is proportional to the light intensity. These photosites actually count the electrons that are produced when photons strike them. The chip takes the place of film in the camera. The rest of the CCD camera comprises electronic devices to record and digitize the signal, plus a cooling system to keep it cool [514]. The electronic video camera or still camera that you have at home contains a small CCD chip, maybe just 1/8-inch square. The Hubble Space Telescope uses one that is quite sophisticated and large. A local amateur observatory here owns an astronomy CCD camera with a chip 1-in. square. As you would expect, cameras with larger chips are more expensive. So far, most astronomy CCD cameras do not have chips as big as 35 mm film negatives, but those size chips are now available. Consequently, they cannot capture a field of view as large as film. However, for planetary imaging, we don't care. A small chip can work well for us.

Astronomy CCD cameras do not have a camera lens. Rather, the camera is fitted with a barrel similar to an eyepiece and is placed into the eyepiece holder of the telescope, much like a 35mm camera body is attached to a telescope for prime focus or eyepiece projection photography. Like a film camera, a CCD camera can be used for long exposures or short ones. Thus, CCD cameras are suited for faint, deep sky work as well as planetary imaging.

As books on CCD cameras will point out, it is important to match pixel size with the image scale produced by your telescope. Once the image is taken, there is then the problem of processing the image in your computer to bring out detail in the finished image. Don Parker points out that, even though the raw CCD image requires further processing, computer work has many advantages over the long hours astrophotographers used to spend in the darkroom with funny lights and chemicals!

Because CCD cameras are so much more sensitive than film, exposure times for planetary imaging are greatly reduced. Instead of several seconds for film, a CCD image can be obtained with exposures of 1 s or less, improving the ability to freeze the image. Most CCD cameras take only black and white images. To obtain color, separate exposures must be taken through red, green, and blue filters. Then the three images must be combined in a computer by using image-processing software. Because these images are going to be combined or stacked, it is important to obtain the three exposures without too much delay between images. Remember, Jupiter rotates rapidly. Therefore, CCD cameras with fast download times are advantageous. With the advent of software that will automatically align and stack images, many CCD imagers today take several images, tens if not hundreds, and stack them together. This stacking and aligning of multiple images accomplishes a couple of important things. Stacking multiple images increases the density of the image, which leads to an image of higher contrast and resolution. Some of the CCD images of Jupiter being produced today by skilled amateurs are of Voyager Spacecraft quality! Absolutely superb imaging showing great detail! Some amateur CCD images have even captured surface features on Jupiter's moons Ganymede and Io.

While CCD cameras are much more affordable today, the really good ones are still quite expensive. An adequate camera will cost at least several hundred dollars, and a really good one will cost several thousand. It is a wonderful research tool for those who can afford one.

9.10 Imaging with Webcams

The advancement of affordable CCD cameras has been a real boost to amateur astronomers who can afford them. And just when it appeared things could not get better, another, much more affordable camera has come along to help the rest of us. The little, inconspicuous webcam, has finally put digital planetary imaging within reach of every amateur astronomer today! Of note, the recent close apparition of Mars was the most imaged apparition of that planet ever, due mainly to the availability of quality webcams. The astronomy world has never seen anything like it! Since a webcam is within affordable reach of almost every amateur today, we will discuss webcam imaging in some detail.

Webcams, when first introduced, were intended for use with computers, to send live communication over the Internet. We could talk with and see our friends and relatives at the same time. The webcam produces live, streaming video images. It wasn't long until some industrious amateurs figured out how to use webcams for astronomy to capture images of bright objects, namely planets. Today, many webcam manufacturers provide software to allow webcams to be used to capture images through a telescope. My personal favorite is the Phillips ToUCam Pro webcam (Fig. 9.9). This camera can be obtained with the proper software and a screw in barrel so the camera can be inserted into the eyepiece holder of a telescope (Fig. 9.10).

The really wonderful news about webcams is that they produce a color image and are extremely affordable, especially when compared to CCD cameras. Mine cost less than $300, including the software and eyepiece barrel! And they are plenty sensitive enough for planetary imaging.

Webcams use a small CCD chip, about 1/4 in. or smaller to collect an image. This image is an avi. file. With the software provided, this stream of images is broken

Fig. 9.9. A Phillips ToU Cam Pro webcam, used to capture images of the moon and planets. (Credit: John W. McAnally).

Fig. 9.10. A webcam mounted into the eyepiece holder of a telescope ready for imaging. A 2× barlow lens is inserted between the telescope and tne camera to increase the focal length to f/20. (Credit: John W. McAnally).

down into individual frames, and the frames are counted. A shutter speed can be specified for the exposure, and gain and contrast can also be adjusted for the camera. The gain and exposure time controls the brightness of the image being captured. I most often use a gain setting of 50–75% and a shutter speed of 1/25 to 1/50 s. With this setup, even my 8-in. telescope has produced surprisingly detailed images of Jupiter.

Fig. 9.11. The authors webcam imaging setup, including webcam, laptop computer, telescope, portable electric power source for the telescope clock drive, and the all important logbook for recording the imaging data. (Credit: John W. McAnally).

With the webcam hooked up to my laptop computer (Fig. 9.11), hundreds of frames can be captured in just a few seconds. It is not unusual for me to capture 500 frames in 30–40 s. Webcams can even be used on really large (Fig. 9.12) telescopes!

The individual raw images reveal a surprising amount of detail, but need to be processed to really bring out the full potential in the final image. The frames must be aligned and stacked. Fortunately, software is available to do just that, and the best part is, the software is free!

There is a variety of software available that will align and stack images. The one I prefer is called *Registax*. This is amazing software, written by Dutch amateur astronomer Cor Berrevoets, and then offered by him to the world for free [515]. What a wonderful gift to the world of astronomy! The software can be obtained from the maker over the Internet and comes with complete instructions. It is very easy to use. The website is *http://aberrator.astronomy.net/registax/*.

To take images with a webcam, set up your telescope as you would for visual observing. While I have seen webcams used successfully with dobsonian mounted reflectors that did not track with the Earth's rotation, I prefer to use an equatorially mounted telescope. I also perform a reasonable polar alignment. However, I do not take as much time aligning as I would for a one-hour deep sky exposure.

Fig. 9.12. The webcam can be used in any size telescope. Here the author installs his webcam into the eyepiece port of a 2-ton, 24-in. (0.61 m) professional classic cassegrain. (Credit: Trudy LeDoux/ the Central Texas Astronomical Society).

A reasonably close alignment is good enough. A webcam can record images that can be stacked with a less than perfectly aligned telescope. However, I prefer a good polar alignment because it will reduce declination drift, allowing the telescope to keep the image centered without frequent adjustment. I prefer the image stay in the center of the field of view at least 10 min before I have to center again. This way, the aligning and stacking program does not have to work so hard making the alignment. Remember, we are trying to capture images that are as crisp as possible. So, exercise a little patience here in setting up the telescope. It is also important to setup the telescope far enough in advance of the imaging run so the telescope can settle down to the outside air temperature. Air currents inside the telescope tube will degrade the images you capture. I often set up my telescope outside prior to having dinner. That usually allows enough time.

Using one of my telescopes, an 8-in., $f/10$ Schmidt-cassegrain, let's go through a typical imaging run. I plug the webcam into my laptop that also runs the imaging software. After setting up the telescope so it can settle down, and polar aligning it, I locate Jupiter in the telescope field of view with a low power eyepiece of about 80×. Then I insert the webcam. Usually, the image will be out of focus. Patience is needed here since the out of focus image will appear faint on the computer screen against the background sky and may be difficult to make out. With a little practice, you will learn which direction to adjust the focus to begin bringing the image into focus on the screen. As soon as I have an image of Jupiter on my computer screen, I use the slow-motion controls on my telescope to finish centering the image. Now I finish focusing. This can be tricky, because touching the telescope to adjust the focus also causes the image to dance on the computer screen. Once I am near focus, I begin moving the focus ever so slightly, past focus one way, then the other, until I have bracketed the focus and I know it is as good as I can make it. It is a good idea to use vibration-damping pads under the telescope's tripod feet when a tripod is used. An electric focusing control can also be a handy tool, since that allows you to avoid touching the telescope. A breeze or wind can also play havoc with an imaging run.

At this point, the telescope is still working at $f/10$. At $f/10$ the image in the camera is big enough to capture the larger features on Jupiter's disk. But, we want even more detail. Now that Jupiter is in focus and centered in the field of view, I remove the camera and insert a 2× Barlow lens into the telescope in front of the camera so that the telescope is working at $f/20$. Now I have a much larger image and even smaller features can be captured. I have often captured the small SSTB ovals with this configuration. Some amateurs take images at f/40 and capture some incredible detail. But remember, the focusing and other issues become more difficult as the image size increases. But it can be worth the effort, especially when the seeing is especially good. Once a sharp focus is achieved, the camera controls can be set. Clicking on "options" allows access to "properties". Click "properties" and the "video properties" screen needed to setup the camera for imaging appears. On the "image control" screen I normally set frame rate to "25", "brightness" to about 25%, "gamma" to about 45%, "saturation" to about 50%, and "auto exposure" to off. On the "camera controls" screen I set "white balance" to auto. Then I set the "shutter speed" between 1/25 and 1/50 s, depending upon the seeing condition, and I set "gain", from low to high at 50–75%. The computer displays a live view of Jupiter on the screen. When these controls are properly set, you can easily make out Jupiter's belts and zones on the computer screen. If you make the image too bright, the image you capture will be washed out and you will not be able to process it properly. Likewise, the image cannot be too faint. These pointers will get you started, but you should experiment to see what settings work best with your telescope.

Now it is time to take the stream of images. After closing out the "video properties" screen, click on "file" and "set capture file" will appear. Clicking "set capture file" allows you to designate the file your image will be saved to. I like to incorporate the date into the file name. This can help you organize your data base of images later. Once the file has been designated you can go back and click on "capture". Then click "start capture". Then click "OK" and the camera will start recording and the frame counter will show the number of images being taken. The video *avi* files will quickly take up a lot of memory. You will want to capture an adequate number of frames so that you have enough to align and stack; however, after a point, additional frames do not gain you very much. I like to take at least 500 frames. You can stop the picture taking process at any time manually by clicking the stop button, or you can set a time limit. I like to set a limit of 30 s. This time limit usually results in about 500 frames. Now I have an *avi* file that is between 300 and 400 MB.

After performing several imaging runs, the images will need to be processed. Sometimes I do this at the telescope while I am waiting for Jupiter to rotate and present a new view. This way, I can also make sure my camera settings are working for the seeing conditions, etc. Or, the images can be processed at any later date, at my convenience. The processing software *Registax* is a wonderful program that is extremely easy to use. The raw images often capture more detail than the observer could draw accurately by hand but they do not reveal the detail that is hidden within them. The *Registax* processing software will work magic on these images!

Registax is an easy program to use. Many of the applications used in this software are the same as those used to align, stack, and process CCD images. The raw *avi* file frames are run through these applications to produce the final detailed image. It is possible to over process images. However, *Registax* allows you to monitor the progress of the processing and steps can be repeated and changed prior to saving the final image. With practice, you can become skilled with the application of this software in a short amount of time.

Once *Registax* is up and running, you will go to the file folder, as with any file, and locate your raw image file. Clicking on "select input" allows you to go to your file folder to select the raw image to be processed. Click on the image file you want to process. Then click "open". Now the raw image appears on the screen. Once *Registax* has the image loaded the processing can begin. You can process all the frames in the *avi* file or you can select a range of images by checking the "Show frame list". Once the list appears, designate the frames you want to include in the processing. Remember, we need enough frames to get a good, dense image with enough contrast and resolution. Normally I set the processing area to 512 pixels. Set the "alignment box pixels" to a number large enough to allow the alignment box to completely surround Jupiter's disk on the screen. Normally I set the alignment box pixels to 256. A white cross will appear on the screen. Move this cross to a feature on Jupiter you can make out, such as a bright spot or a festoon; then click the left button on the mouse. This will cause the "Aligning" screen to appear. Now you have successfully designated an alignment reference point on Jupiter's disk and you are ready to set up the alignment parameters. An FFT spectrum box will appear along with a "registration properties" box will appear along with the image screen. Also, various values will appear on the "aligning" screen that you will need to set. Under "Optimizing options" I normally set "Optimize until" to 1%, "Search area" to 2 pixels, and "Lower quality" to 80%. Under "Tracking options" I check "Track object". Under the other "Options" I check the "Alignment filter" and set it to 5 pixels. I set "Quality filter band" to start 2 and width 5. I check "use contrast", and I set "resample" to 2.0 and "Bell". I check "Auto-Optimization" and "Fast optimize". The other settings I leave blank. Now you are ready to align your images. Click the "align" button. Once the frames are aligned the program will "optimize" them. The "registration properties" box display will show the process of this step and indicate the frame order with best quality first and indicate graphically how close the alignment actually is. It takes my laptop about 12 min to align and optimize 550 frames with the settings I use. This aligning process does what is says; it takes all of the frames you have selected and, regardless of where they appeared in the camera's field of view, the software "picks them up and moves them so they are in the same position on each frame", using features seen on the planet itself to make the alignment. At least, that is how I visualize the process. After the aligning process is completed, the stacking process can be started.

To start the stacking process, just click on the tab labeled "Stacking" at the top of the computer screen. The stacking page will appear, displaying Jupiter's image along with a "Stackgraph". The stacking process does just what it says, it stacks or digitally adds all the images together. I think you begin to see the benefit of the procedures the *Registax* software is performing. Aligning and stacking multiple images will average out imperfections in the individual images. Plus, the weak color and contrast in single images is improved by adding several images together.

The "Stackgraph" is used to select the frames that will be stacked together. This graph represents the frames in order of quality, indicating high to low with a horizontal red line running across the top of the graph. This graph line tells you where the quality of the individual frames falls off. Eliminating the lower quality frames from the stack can improve the final finished image. Moving a "cutoff" bar across the graph left to right makes this selection. A blue line that looks like an EKG indicates registration or alignment differences in the images. A horizontal blue line can be moved vertically to set the limit for alignment quality; again, improving the result in the finished image. In other words, we only want the best frames, both in quality and alignment, to end up in our finished image. You will quickly become comfortable with this part of the process as you gain experience with the software. Stacking quality percentage and registration difference percentage is indicated at the bottom of the screen. Normally, the program will default to selecting the best 101 frames to stack. I normally do the stack initially with these 101 frames and see how I like it. I can always come right back and repeat this part of the process if I see I need more frames to get an acceptable finished image.

Once the graph is set to your satisfaction, you can start the stacking process by clicking on the "stack" button. The stacking process begins and a "progress" graph appears, counting down the stacking percentage completed. For 101 frames, this takes about 2 min. When the stacking process is completed, a greatly improved image will appear on the screen. But, we are still not finished. There is one more process called "wavelet processing." Clicking on the "wavelet processing" tab at the top of the screen moves us to this next step.

To me, wavelet processing is the really "magical" part of image processing! This final step magically pulls out, or sharpens, the detail that is hidden in the image. Manipulating slides to adjust the "layer" settings performs this sharpening. You will see the image "sharpen up" as each layer is adjusted so it is easy to visually judge the effectiveness of your processing. This is a valuable feature of the software since it is easy to over process during this phase. If not satisfied with the finished image, you can easily go back and redo the stacking process and the wavelet process. Once you are satisfied with the finished image you can save it by clicking on the "save" button near the bottom right corner of the screen. A box will pop up that will allow you to give the image a filename and designate a folder. Click "save" in this box and the process is complete. For further reference, I recommend you read the magazine article about Registax, written by the creator Cor Berrevoets, which appeared in the April 2004 issue of *Sky & Telescope* magazine. It is a more detailed discussion of what I have presented here. I understand that Registax version 4 is now available and I have had really good reports about it.

Of course, the webcam and Registax software can be used on the moon and other planets as well. You will soon realize, as thousands of webcam users have already figured out, that the webcam is fantastic! It is much easier and much faster to use than a conventional astronomy CCD camera, and it is super affordable!

9.11 Making use of CCD and Webcam Images

Once you become comfortable in the use of your CCD camera or webcam, it will be no time at all until you will have collected a number of beautiful images. There can be a great deal of personal satisfaction in taking and processing the perfect image as the artistic side of imaging is certainly part of the pleasure. But, if you want your images to be more than just pretty pictures, then you have to be a scientist and make them mean something. Of course, that is the major focus of this book, to encourage the reader to study Jupiter and add something to our knowledge of the planet. Certainly, your images will become part of the historical record of the physical appearance of the planet; but there is more to be gleaned from these images. For your images to be truly worth something scientifically, they will need to be measured to determine the position of features seen on Jupiter's disk.

Previously we discussed the use of central meridian transit timings to determine the longitudinal position of features seen on Jupiter's disk. Measuring CCD and webcam images has the same goal, to keep track of features over time. The longitude of features can be scaled from the planet's disk, using the images from webcams and CCD cameras.

Using a fan shaped device described in Rogers [516] (Fig. 9.13), it is relatively easy to determine longitude of any feature seen on the disk if the longitude of the central meridian is known for Jupiter at the time the image was taken. Once again, it is important to record all the pertinent information regarding the image just like we do for a disk drawing or transit timing. The imager should record the date and time to the nearest minute in Universal Time, the telescope size and focal ratio, seeing conditions, transparency, and the location of the observer. Additionally, information should be recorded regarding image processing, including the camera used, filters if any, integration times, number of frames and frame speed, and any other information that will help the user of your images understand how they were produced. This data gives credibility to your image and makes it useful. For example, without the date and time the image is almost useless for science.

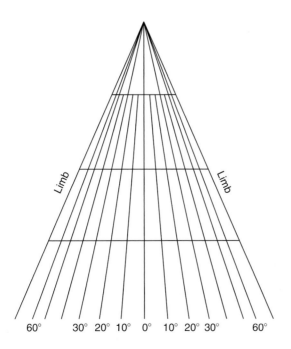

Fig. 9.13. A measuring device used to scale the longitude of features on a CCD or webcam image. (Credit: John Rogers, 1995).

Knowing the central meridian of Jupiter in your image, Roger's device is used to scale the longitude of features seen on either side of the central meridian. The image can be examined on your computer screen or you can print it out onto paper. To use the device, place it onto Jupiter's image with the outer rays of the fan on the left and right limb of the planet at the latitude of the belt or zone containing the feature you wish to scale. With the disk of Jupiter oriented so south is at the top, place the fan over Jupiter so that the widest opening of the fan is up if measuring features in Jupiter's northern hemisphere, and place it with the widest part of the fan down if measuring Jupiter's southern hemisphere. Each division in the fan represents 10° of longitude. With Jupiter oriented so that South is up, longitude on Jupiter will increase from left to right, or from the preceding side to the following side. So, a feature that is to the left of Jupiter's central meridian will have a lower longitude than the central meridian, and so on. I have used this device with great success, achieving an accuracy of one to two degrees with my measurements. Measuring is not difficult; it just requires patience and attention to detail. Even old film photographs of Jupiter can be measured using the Rogers device as long as you know what the central meridian is for Jupiter on that photo.

The longitudinal positions of features should be recorded on the transit form previously discussed, in the same manner as a central meridian transit timing using the same nomenclature to describe the objects measured. We have already discussed the value of recording the longitudinal position of features over time. This is one of the most important contributions the amateur astronomer can make.

CCD and webcam images have a great advantage over central meridian transit timings in that the images are not subjective and the measurements are not prone to timing errors, as transit timings can be. Images can also be used as a record of color and intensity, but care must be taken not to over process or to increase the contrast too much during processing. Color can also be skewed if care is not taken in the application of filters and color processing. Imagers should strive for a finished image that shows great detail but that is as natural as possible. One imager that has excelled at this is Ed Grafton of Houston, Texas. Much can be learned from his work. It is exceptional. To keep myself honest in my own image processing, I make regular visual observations of Jupiter so I know how the planet is supposed to look. Besides, as exciting as imaging can be, still nothing compares to seeing the planet visually in the eyepiece of a telescope.

9.12 Measurement of Latitude

As part of the long-term study of Jupiter, the observation of latitude is also important to our understanding of the behavior of the planet. We know from measurements over time what the normal latitude boundaries of Jupiter's belts and zones are, we know the normal latitude of Jupiter's currents and jet streams, and we know that these latitudinal positions can vary. While the measurement of latitude is not considered by many observers to be as easily performed as measurements of longitude, it is an area of study that amateurs are certainly capable of participating in.

During the 1800s and until about 1950, measurements of latitude could only be made visually with telescopes equipped with filar micrometers. These were precision measuring devices that were very expensive and not often available to amateur astronomers. The procedure for using a filar micrometer is fully discussed in Peek [517]. In order for filar micrometer latitude measurements to be useful, great care and precision must be exercised in making the measurement. Furthermore, the telescope used must be equatorially mounted on a very stable mounting that is driven to compensate for the Earth's rotation. Otherwise, the image in the eyepiece will bounce around hopelessly, making an accurate measurement impossible and unreliable.

Fortunately, the procedure for making latitude measurements can also be applied to high-resolution photographs, and this has been the case since about 1948 [518]. Photographs of high resolution are required, since any blurring of the image makes it difficult to make measurements with precision.

The advent of CCD and webcam imaging has finally made it possible for astronomers, including amateurs, to make measurements of latitude from images with true accuracy. Once again, CCD cameras and webcams are making it possible for amateurs to perform scientific work that is of professional quality. Today I am not aware of very many amateurs who use filar micrometers.

Once the raw measurements have been taken, usually from images today, they must be reduced. On Earth, we refer to latitude as geographical or geocentric. On Jupiter, the corresponding latitudes are named 'zenographical' and 'zenocentric' after the Ionic genitive of the Greek 'Zeus' [519]. Some astronomers use the terms 'planetographical' and 'planetocentric'. Peek [520], Rogers [521], and Schmude [523] discuss the methods for doing this. In order to calculate the latitude from measurements, the following steps must be performed (Fig. 9.14). Measure the distance from the feature from the south pole, s, and the polar radius, r. Calculate $(r-s)/r$. Call this $\sin\theta$, and solve for θ. The next thing we have to determine is Jupiter's apparent tilt, which we designate δ'. Most almanacs give the Earth's zenocentric latitude δ (or D_E), but we need Jupiter's tilt, and so we must solve for δ'. δ' is derived by: $\tan\delta' = 1.07\tan\delta$. (1.07 is the ratio of Jupiter's polar and equatorial diameters, and is recently actually determined to be 1.0694.) Now we can calculate the "mean latitude" β', which is: $\beta' = \theta + \delta'$. To summarize:

$$\beta' = \sin^{-1}\frac{(r-s)}{r} + \delta'.$$

Because Jupiter is oblate and not a perfect sphere, there are two other definitions of latitude, and β' will need to be converted into one of them (Fig. 9.14). Zenocentric latitude (β) is the angle subtended at the planet's center between the feature and

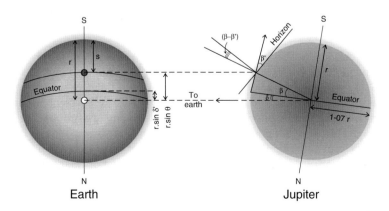

Fig. 9.14. A diagram illustrating the measurement of latitude. (Credit: John Rogers, 1995).

the equator. Zenographic latitude (β'') is the angle between the polar axis and the local horizon.

We can convert between the two having determined their tangents:

$$\tan \beta'' = 1.07 \tan \beta' \quad \text{or} \quad \tan \beta' = 1.07 \tan \beta''.$$

If Jupiter were a perfect sphere, β' and β'' would be identical, but on Jupiter they are not and can differ by as much as 4°.

For the purpose of making latitude measurements of Jupiter's features, we normally want to determine latitude like we would perceive it on Jupiter's "surface". Thus, we normally want to determine zenographic latitude, β''.

Richard Schmude [524] expresses these calculations in a slightly different mathematical form. First compute θ, as:

$$\text{Sin } \theta = (s-n) / (s + n)$$

where s and n are the distances between the feature and the south and north pole at the polar limb. These distances are measured with a ruler in millimeters. If θ is negative, then the feature is south of the equator, and if θ is positive, the feature is north of the equator. Since θ represents latitude on Jupiter as though Jupiter were spherical, we must solve for the other calculations of latitude using the following equations:

$$\text{inv } \tan\left[1.0694 \times \tan\left[\text{inv.sin } (\theta) + 1.0694D\right]\right] \text{ zenographic}$$

$$\text{inv } \tan\left[1.9351 \times \tan\left[\text{inv.sin } (\theta) + 1.0694D\right]\right] \text{ zenographic}$$

$$\text{inv sin } \theta + 1.0694D \text{ mean,}$$

where D is the sub-Earth latitude on Jupiter. This quantity D can be found in the Astronomical Almanac [525].

In order to perform accurate measurements of latitude, the images used must be of high resolution. Images in which the boundaries of belts and zones are ill defined, or in which bright and dark spots appear indistinct, will result in measurements of which the margin of error is too great, as an accuracy of measurement within one degree is desired. The amateur astronomer who can make latitude measurements with the desired accuracy can certainly consider himself to be a more advanced amateur.

9.13 Keeping a Record: The Logbook

Most of us find it relatively easy to remember the date and circumstance surrounding the individual pictures taken over the years in the family photo album. Family pictures are, well, familiar to us. We know the people in them, or we remember in great detail the family vacation they depict. Family pictures represent familiar surroundings. The collecting of scientific data, or the making of astronomical images or sketches is a totally different situation. It is not nearly so easy to remember the circumstance surrounding these; and frankly, it is not very scientific or reliable to work from memory. As emphasized previously, a drawing or image for which no basic information is recorded is practically worthless.

We should record basic information regarding date, time, instrument used, and weather conditions for each observation of course; but, it is also most beneficial to assemble this data in some logical, organized manner.

For me, one of the best ways to keep an organized record is the use of a logbook. Keeping as logbook is a natural thing to do. I have kept one since 1967. My logbook is not a fancy, leather bound book; it is a simple, spiral bound notebook. I have filled up several of these over the years, and when one is completed it is relatively easy to go buy another one and keep going.

I use my logbook for all of my observations, not just the planetary observations. And so for me, it is also a nice history of what I have been up to. When I am observing Jupiter or any planet, I start the entry for the night with very basic information. This includes the day of the week, the date in local time, and my location. Yes, I enter the local date, and I note the time zone and whether it is standard time or daylight saving time. Later, I will note the date in universal time. I use local time along with universal time because I like to look back and be reminded of the local conditions, the time of night, the conditions, and so on. I also make a general remark about the weather conditions, cold or hot, windy or still, light haze or clear, and whether our moon is bright or absent. I also record when an observation previously anticipated could not be performed because the weather deteriorated. Some of these notations are more nostalgic than scientific, but I like to record them. Truthfully, you can record anything you want to. I usually mention the names of people who may have observed with me, such as friends or family. And, I have on occasion asked distinguished friends and acquaintances to autograph my logbook on that night's pages. I also record what my intentions are for the night, for example to observe Jupiter. As you can see, my logbook is a very personal record. It is a record containing not only data that was carefully and scientifically recorded, but also one that I enjoy going back to and reading over again. There are many pleasant memories in my logbooks.

After the general comments, I record more serious information, such as the type and size of telescope to be used and camera equipment, if any. For Jupiter, before I begin a sketch or imaging run, I take the time to observe visually for several minutes. I then enter into my logbook a description of Jupiter's general appearance. At this time I may also make and record my intensity estimates of Jupiter's belts, zones, and other features. I also make note of the seeing and transparency, and the magnification and filters used for this phase of my observation. Finally, I am ready to proceed with the more demanding observations of the evening. Each type of observation, whether visual transit timings or webcam imaging, has its own set of information that needs to be recorded in a careful, systematic manner.

For visual transit timings, the very best example of data recording is the typical notebook entry made by Walter Haas. Walter Haas is the founder of the Association of Lunar and Planetary Observers and has been an indefatigable observer of Jupiter. I remember one apparition in which Walter recorded over 1,600 individual transit timings! My notebook entries for Jupiter are modeled after Walter Haas. First I enter information regarding the size and type of telescope used and the observing location, such as city and state, etc. Then, working from left to right, I set up the rest of my notebook page to record the following information:

1. The number of my transit timing. (I keep track of my transit timings by numbering them in order from the beginning of the apparition to the end of the apparition. I do not start the numbering over each night.)

2. A description of the feature observed in transit, in the notation previously discussed.

3. The time of the transit, to the nearest minute, in universal time.

4. The longitude, usually determined later by referencing an ephemeris, with the column subdivided into two smaller columns, one for System I, and the other for System II.
5. And finally I record information regarding the magnification and filters used, seeing, and transparency.

After I am finished recording transits for the night, I record any other descriptive notes that I feel will further describe the observations.

Imaging runs with a CCD camera or webcam require their own special data. For webcam imaging, I enter the following information, from left to right:

1. The image number, in order, for the night.
2. The end time, in universal time, for the imaging run.
3. The number of frames accumulated.
4. Total elapsed time for the imaging run.
5. The exposure setting.
6. The gain setting.
7. The *f*/ratio setup; *f*/10, *f*/20, etc.
8. The seeing.
9. The transparency.

I leave space between these entries so I can enter any other information I feel is necessary; and also so I can come back later and write a comment regarding how the finished image came out, after *Registax* processing. I also leave room to note the file name.

The use of the logbook is plain common sense. Once you record the necessary data for each type of observation, you can record anything you want or find interesting. It can be the journal of your astronomical life. Get into the habit of always keeping a logbook. Since you are taking the time to go out with equipment and look at the planets or stars anyway, you may as well make a record of it. You never know when the insignificant observation made the night before turns out to be something of great importance, and you are the only one that saw it.

9.14 Reporting

The goal of scientific research is publication. The work of the most brilliant scientist will come to nothing if others do not know about it. There are professional publications, journals that regularly publish the works of professional scientists. These journals are discriminating publications that require peer review and acceptance by a committee of scientists before the research work can be accepted for publication. The point is, getting work published is a difficult thing, and it is meant to be. The reputation of the journal, the institution, not to mention the scientist and the science itself, is at risk and can be irreparably harmed by bad science. Professional scientists make their living doing their science. Their careers depend upon acceptance and publication. But at least professional scientists have publications in which they can be published.

Amateurs also have venues in which their work can be recognized. The work of some amateur astronomers is of such high quality that professional astronomers will often incorporate it into their own research. The amateur may be honored, in some cases, by having his name listed with the authors of the publication. Or, the amateur may be credited in the dedication segment of the book or publication, or in some other manner. The American Astronomical Society (AAS), an organization that limits its membership to professional astronomers or other scientists in closely related fields, has undertaken an effort to foster collaboration between the professional and amateur astronomy community, recognizing the potential of amateurs to contribute to science. The Division for Planetary Scientists, a sub-group of the AAS, has long recognized the potential of amateurs to contribute to the work they do. There have been many occasions in which professional papers on Jupiter have incorporated data gathered by members of the British Astronomical Association (BAA), the Association of Lunar and Planetary Observers (A.L.P.O.), and other organizations. I myself have received telephone calls and e-mail messages from scientists seeking transit data I have accumulated for A.L.P.O. These are certainly exhilarating experiences for any amateur scientist.

In a sense, amateurs have a great advantage over the professional astronomer. We can work at our own pace. We can observe with our own instruments anytime we choose, not having to wait for telescope time. Most of us make our living in other jobs and professions, so we are not encumbered with fund raising to do our science. In short, amateurs have a lot of freedom. If we want to do astronomy just for fun, we can!

But astronomy is the one remaining science that also offers the amateur a chance to make a meaningful contribution. If the amateur wants to do serious science, he can. As we have seen, improvements in equipment and affordability are making it possible for the serious amateur to do many of the things that used to be the realm of the professional astronomer.

We have already discussed the kinds of observations that amateurs can perform, many of which require only modest equipment and attention to detail. So when the amateur has collected his data, what does he do with it?

9.15 Amateur Organizations

Today in this hobby and science of astronomy, and especially in planetary observing, we are blessed with a number of very good, accessible amateur organizations that cultivate amateurs and provide a place for them to learn and share their observations. Many organizations have web sites with practice aids, discussion groups, and leaders available to help the beginner launch an observational career, whether the career is intended to be casual or serious. Information regarding various organizations will be provided in Chap. 9 of this book. You should not fail to explore these organizations. There are also many astronomy societies and clubs and you may be fortunate to find one in your locale. Joining a club and visiting with its members provides an excellent opportunity to learn more about all aspects of astronomy. I have always found club members more than willing to share their special knowledge with the beginner.

Surprisingly, there are many amateur organizations that are eager to receive the data and observations collected by careful amateurs. Many of these organizations also have training programs that will help the novice who wants to become more serious about his science. The Association of Lunar and Planetary Observers, the British Astronomical Association, and other organizations in Europe and the Orient all provide venues for the amateur to contribute and publish his work. These organizations, although composed of amateurs, maintain high standards of work. The reputation of these amateur organizations is as valuable to them as the professional community's is to it. And knowing this, professionals often turn to these amateur organizations for assistance. During a recent Jupiter apparition, professional astronomers called upon the amateur community to conduct world wide, round the clock imaging of Jupiter during opposition, so that a complete record of Jupiter's physical appearance could be made. This mission was very successful, and united the amateur and professional communities like nothing else could.

Amateur organizations maintain their own publications, and seek to publish on a regular schedule. Almost any amateur who does good science can submit his work. The A.L.P.O. publishes the Journal of the Association of Lunar & Planetary Observers *The Strolling Astronomer* on a quarterly basis. The BAA and other organizations have similar publications. The work of these organizations is impressive and very professional.

Serious amateur organizations take great pride in what they do and zealously guard their reputations. These amateur organizations are typically membership organizations whose members do not participate in astronomy as a way of making a living, but rather participate for the love of the hobby and science. Yet, they want their work to be taken seriously and strive for the highest possible standards in their observations and data. The standards must be high if their work is to amount to anything. Although there are many high quality organizations around the world, I am most familiar with two of them, the British Astronomical Association (BAA) and the Association of Lunar and Planetary Observers (A.L.P.O.).

The contact information for the BAA is:

British Astronomical association
Burlington House
London
UK
W1J 0DU
http://www.britastro.org

The contact information for the A.L.P.O. is:

Matthew Will
A.L.P.O. Membership Secretary
P.O. Box 13456
Springfield, Illinois 62791–3456
http://www.lpl.arizona.edu/alpo

Certainly, we can and often do participate in astronomy just for fun, just for the beautiful sights we see in a telescope. But think about this; if you are going to take the time to haul all of that equipment outdoors anyway, why not do something serious with it. Make an observation that means something, record it properly, and report it to the appropriate organization. It could be you making the next meaningful contribution to this science of astronomy.

Conclusion

What do We do Now?

A college professor I knew, speaking to a group of college students upon the occasion of their graduation, remarked, "Graduates, I have taught you everything you know; but, I have not taught you everything I know! Your work has only begun!"

How true that statement is for all of us. If you are new to observing Jupiter, what you have just read in this book may comprise your total knowledge about the planet. But your learning experience has only just begun. There is so much more to learn and new discoveries are waiting for someone to discover them! I think that is where the fun is; just a few more degrees of rotation, a few more degrees of longitude and new, different, ever changing features wait to be seen and recorded. Tonight, tomorrow night, next week, next month, next year; what does the future hold? Only a practiced observer will be able to discern the changing surface of Jupiter!

The observations of amateurs will continue to be important in maintaining the history of the physical appearance of Jupiter. At present there is no spacecraft in orbit around the planet. The New Horizons Spacecraft took images of Jupiter as it flew past a few months ago on its way out to Pluto, but now it too is beyond the giant planet. So once again only Earth based instruments will be able to keep watch. Now it is up to us.

Whether you observe for fun or seek to make a serious contribution, the pastime of Jupiter watching offers something for everyone. So, get off the couch, grab your telescope and head outside. Your adventure starts at sundown!

References

1. *The Giant Planet Jupiter*, J. H. Rogers, Cambridge University Press 1995, p. 6
2. *The Giant Planet Jupiter*, J. H. Rogers, Cambridge University Press 1995, p. 4
3. *The Planet Jupiter*, B. M. Peek, Faber and Faber, Ltd., 2nd edition Patrick Moore 1981, p. 83
4. *The Giant Planet Jupiter*, J. H. Rogers, Cambridge University Press 1995, p. 105
5. *The Planet Jupiter*, B. M. Peek, Faber and Faber, Ltd., 2nd edition Patrick Moore 1981, p. 83
6. *The Giant Planet Jupiter*, J. H. Rogers, Cambridge University Press 1995, p. 106
7. *The Giant Planet Jupiter*, J. H. Rogers, Cambridge University Press 1995, p. 106
8. *The Giant Planet Jupiter*, J. H. Rogers, Cambridge University Press 1995, p. 116
9. *The Giant Planet Jupiter*, J. H. Rogers, Cambridge University Press 1995, p. 118
10. *The Giant Planet Jupiter*, J. H. Rogers, Cambridge University Press 1995, p. 118
11. *The Giant Planet Jupiter*, J. H. Rogers, Cambridge University Press 1995, p. 123
12. The Giant Planet Jupiter, J. H. Rogers, Cambridge University Press 1995, p. 127
13. *The Planet Jupiter*, B. M. Peek, Faber and Faber, Ltd., 2nd edition Patrick Moore 1981, p. 88
14. *The Planet Jupiter*, B. M. Peek, Faber and Faber, Ltd., 2nd edition Patrick Moore 1981, p. 94
15. *The Planet Jupiter*, B. M. Peek, Faber and Faber, Ltd., 2nd edition Patrick Moore 1981, p. 94
16. *The Planet Jupiter*, B. M. Peek, Faber and Faber, Ltd., 2nd edition Patrick Moore 1981, p. 94
17. *The Planet Jupiter*, B. M. Peek, Faber and Faber, Ltd., 2nd edition Patrick Moore 1981, p. 95
18. *Jupiter Odyssey: the Story of NASA's Galileo Mission*, D. M. Harland, Springer/Praxis 2000, pp. 121,122
19. *The Giant Planet Jupiter*, J. H. Rogers, Cambridge University Press 1995, p. 160
20. *The Giant Planet Jupiter*, J. H. Rogers, Cambridge University Press 1995, p. 197
21. *The Giant Planet Jupiter*, J. H. Rogers, Cambridge University Press 1995, p. 160
22. *The Giant Planet Jupiter*, J. H. Rogers, Cambridge University Press 1995, p. 205
23. "New Observational Results Concerning Jupiter's Great Red Spot", A. A. Simon-Miller, P. J. Gierasch, R. F. Beebe, B. Conrath, F. M. Flasar, R. K. Achterburg, and Cassini CIRS Team. *Icarus* 2002, 158, p. 254
24. "New Observational Results Concerning Jupiter's Great Red Spot", A. A. Simon-Miller, P. J. Gierasch, R. F. Beebe, B. Conrath, F. M. Flasar, R. K. Achterburg, and Cassini CIRS Team. *Icarus* 2002, 158, p. 250
25. "New Observational Results Concerning Jupiter's Great Red Spot", A. A. Simon-Miller, P. J. Gierasch, R. F. Beebe, B. Conrath, F. M. Flasar, R. K. Achterburg, and Cassini CIRS Team. *Icarus* 2002, 158, p. 250
26. *The Giant Planet Jupiter*, J. H. Rogers, Cambridge University Press 1995, p. 188.
27. "New Observational Results Concerning Jupiter's Great Red Spot", A. A. Simon-Miller, P. J. Gierasch, R. F. Beebe, B. Conrath, F. M. Flasar, R. K. Achterburg, and Cassini CIRS Team. *Icarus* 2002, 158, p. 249
28. "New Observational Results Concerning Jupiter's Great Red Spot", A. A. Simon-Miller, P. J. Gierasch, R. F. Beebe, B. Conrath, F. M. Flasar, R. K. Achterburg, and Cassini CIRS Team. *Icarus* 2002, 158, pp. 249, 250
29. *The Giant Planet Jupiter*, J. H. Rogers, Cambridge University Press 1995, p. 189
30. "New Observational Results Concerning Jupiter's Great Red Spot", A. A. Simon-Miller, P. J. Gierasch, R. F. Beebe, B. Conrath, F. M. Flasar, R. K. Achterburg, and Cassini CIRS Team. *Icarus* 2002, 158, p. 250
31. "New Observational Results Concerning Jupiter's Great Red Spot", A. A. Simon-Miller, P. J. Gierasch, R. F. Beebe, B. Conrath, F. M. Flasar, R. K. Achterburg, and Cassini CIRS Team. *Icarus* 2002, 158, p. 250

32. "New Observational Results Concerning Jupiter's Great Red Spot", A. A. Simon-Miller, P. J. Gierasch, R. F. Beebe, B. Conrath, F. M. Flasar, R. K. Achterburg, and Cassini CIRS Team. *Icarus* 2002, 158, p. 250

33. *The Planet Jupiter*, B. M. Peek, Faber and Faber, Ltd., 2nd edition Patrick Moore 1981, pp. 138, 139

34. "New Observational Results Concerning Jupiter's Great Red Spot", A. A. Simon-Miller, P. J. Gierasch, R. F. Beebe, B. Conrath, F. M. Flasar, R. K. Achterburg, and Cassini CIRS Team. *Icarus* 2002, 158, p. 251

35. "New Observational Results Concerning Jupiter's Great Red Spot", A. A. Simon-Miller, P. J. Gierasch, R. F. Beebe, B. Conrath, F. M. Flasar, R. K. Achterburg, and Cassini CIRS Team. *Icarus* 2002, 158, p. 250

36. R. Schmude, JALPO Spring 2003, p. 26

37. P. Budine, JALPO February 1971, p. 198

38. R. Schmude, JALPO Spring 2003, p. 41

39. "New Observational Results Concerning Jupiter's Great Red Spot", A. A. Simon-Miller, P. J. Gierasch, R. F. Beebe, B. Conrath, F. M. Flasar, R. K. Achterburg, and Cassini CIRS Team. *Icarus* 2002, 158, p. 250

40. *The Giant Planet Jupiter*, J. H. Rogers, Cambridge University Press 1995, p. 191

41. "New Observational Results Concerning Jupiter's Great Red Spot", A. A. Simon-Miller, P. J. Gierasch, R. F. Beebe, B. Conrath, F. M. Flasar, R. K. Achterburg, and Cassini CIRS Team. *Icarus* 2002, 158, p. 251

42. "New Observational Results Concerning Jupiter's Great Red Spot", A. A. Simon-Miller, P. J. Gierasch, R. F. Beebe, B. Conrath, F. M. Flasar, R. K. Achterburg, and Cassini CIRS Team. *Icarus* 2002, 158, p. 251

43. "New Observational Results Concerning Jupiter's Great Red Spot", A. A. Simon-Miller, P. J. Gierasch, R. F. Beebe, B. Conrath, F. M. Flasar, R. K. Achterburg, and Cassini CIRS Team. *Icarus* 2002, 158, p. 264

44. "New Observational Results Concerning Jupiter's Great Red Spot", A. A. Simon-Miller, P. J. Gierasch, R. F. Beebe, B. Conrath, F. M. Flasar, R. K. Achterburg, and Cassini CIRS Team. *Icarus* 2002, 158, p. 253

45. E. Reese, JALPO November 1963, p. 142

46. P. Budine. JALPO May 1974, p. 230

47. Favero et al. JALPO July 1978, p. 97

48. P. Budine, JALPO June 1973, p. 83

49. P. Budine, JALPO January 1986, p. 97

50. P. Budine, JALPO February 1990, p. 2

51. P. Budine, JALPO August 1997, p. 164

52. R. Schmude, JALPO July 1989, p. 124

53. D. Lehman et al. JALPO April 1999, p. 52

54. R. Schmude, JALPO September 1991, p. 113

55. D. Lehman and J. W. McAnally, JALPO October 1998, p. 164

56. "New Observational Results Concerning Jupiter's Great Red Spot", A. A. Simon-Miller, P. J. Gierasch, R. F. Beebe, B. Conrath, F. M. Flasar, R. K. Achterburg, and Cassini CIRS Team. *Icarus* 2002, 158, p. 263

57. "New Observational Results Concerning Jupiter's Great Red Spot", A. A. Simon-Miller, P. J. Gierasch, R. F. Beebe, B. Conrath, F. M. Flasar, R. K. Achterburg, and Cassini CIRS Team. *Icarus* 2002, 158, p. 263

58. "New Observational Results Concerning Jupiter's Great Red Spot", A. A. Simon-Miller, P. J. Gierasch, R. F. Beebe, B. Conrath, F. M. Flasar, R. K. Achterburg, and Cassini CIRS Team. *Icarus* 2002, 158, p. 249

59. *The Giant Planet Jupiter*, J. H. Rogers, Cambridge University Press 1995, p. 220

60. J. W. McAnally, JALPO July 1999, pp. 108–111

61. J. W. McAnally, JALPO July 1999, pp. 108–111

62. *The Giant Planet Jupiter*, J. H. Rogers, Cambridge University Press 1995, p. 220

63. *The Giant Planet Jupiter*, J. H. Rogers, Cambridge University Press 1995, p. 223

64. "The Merger of Two Giant Anticyclones in the Atmosphere of Jupiter", A. Sanchez-Lavega, G. S. Orton, R. Morales, J. Lecacheux, F. Colas, B. Fisher, P. Fukumura-Sawada, W. Golisch, D. Griep, C. Kaminski, K. Baines, K. Rages, and R. West. *Icarus* 2001, 149, p. 492

65. *The Giant Planet Jupiter*, J. H. Rogers, Cambridge University Press 1995, p. 223

66. "The Merger of Two Giant Anticyclones in the Atmosphere of Jupiter", A. Sanchez-Lavega, G. S. Orton, R. Morales, J. Lecacheux, F. Colas, B. Fisher, P. Fukumura-Sawada, W. Golisch, D. Griep, C. Kaminski, K. Baines, K. Rages, and R. West. *Icarus* 2001, 149, p. 492.

67. J. W. McAnally, unpublished report, July 2000

68. "The Merger of Two Giant Anticyclones in the Atmosphere of Jupiter", A. Sanchez-Lavega, G. S. Orton, R. Morales, J. Lecacheux, F. Colas, B. Fisher, P. Fukumura-Sawada, W. Golisch, D. Griep, C. Kaminski, K. Baines, K. Rages, and R. West. *Icarus* 2001, 149, p. 492

69. "The Merger of Two Giant Anticyclones in the Atmosphere of Jupiter", A. Sanchez-Lavega, G. S. Orton, R. Morales, J. Lecacheux, F. Colas, B. Fisher, P. Fukumura-Sawada, W. Golisch, D. Griep, C. Kaminski, K. Baines, K. Rages, and R. West. *Icarus* 2001, 149, p. 492

70. "The Merger of Two Giant Anticyclones in the Atmosphere of Jupiter", A. Sanchez-Lavega, G. S. Orton, R. Morales, J. Lecacheux, F. Colas, B. Fisher, P. Fukumura-Sawada, W. Golisch, D. Griep, C. Kaminski, K. Baines, K. Rages, and R. West. *Icarus* 2001, 149, pp. 492–494

71. "The Merger of Two Giant Anticyclones in the Atmosphere of Jupiter", A. Sanchez-Lavega, G. S. Orton, R. Morales, J. Lecacheux, F. Colas, B. Fisher, P. Fukumura-Sawada, W. Golisch, D. Griep, C. Kaminski, K. Baines, K. Rages, and R. West. *Icarus* 2001, 149, p. 494

72. "The Merger of Two Giant Anticyclones in the Atmosphere of Jupiter", A. Sanchez-Lavega, G. S. Orton, R. Morales, J. Lecacheux, F. Colas, B. Fisher, P. Fukumura-Sawada, W. Golisch, D. Griep, C. Kaminski, K. Baines, K. Rages, and R. West. *Icarus* 2001, 149, p. 494

73. "The Merger of Two Giant Anticyclones in the Atmosphere of Jupiter", A. Sanchez-Lavega, G. S. Orton, R. Morales, J. Lecacheux, F. Colas, B. Fisher, P. Fukumura-Sawada, W. Golisch, D. Griep, C. Kaminski, K. Baines, K. Rages, and R. West. *Icarus* 2001, 149, p. 494

74. J. W. McAnally, unpublished report, July 2000

75. "The Merger of Two Giant Anticyclones in the Atmosphere of Jupiter", A. Sanchez-Lavega, G. S. Orton, R. Morales, J. Lecacheux, F. Colas, B. Fisher, P. Fukumura-Sawada, W. Golisch, D. Griep, C. Kaminski, K. Baines, K. Rages, and R. West. *Icarus* 2001, 149, p. 494

76. *The Giant Planet Jupiter*, J. H. Rogers, Cambridge University Press 1995, p. 224

77. *The Giant Planet Jupiter*, J. H. Rogers, Cambridge University Press 1995, p. 45

78. *The Giant Planet Jupiter*, J. H. Rogers, Cambridge University Press 1995, p. 45

79. *The Giant Planet Jupiter*, J. H. Rogers, Cambridge University Press 1995, p. 48

80. *The Giant Planet Jupiter*, J. H. Rogers, Cambridge University Press 1995, p. 48

81. *The Giant Planet Jupiter*, J. H. Rogers, Cambridge University Press 1995, p. 48

82. *The Giant Planet Jupiter*, J. H. Rogers, Cambridge University Press 1995, p. 49

83. *The Giant Planet Jupiter*, J. H. Rogers, Cambridge University Press 1995, p. 49

84. *The Planet Jupiter*, B. M. Peek, Faber and Faber, Ltd., 2nd edition Patrick Moore 1981, p. 43

85. *The Giant Planet Jupiter*, J. H. Rogers, Cambridge University Press 1995, p. 59

86. *The Planet Jupiter*, B. M. Peek, Faber and Faber, Ltd., 2nd edition Patrick Moore 1981, pp. 41, 42

87. *Visual Observations of the Planet Saturn and its Satellites: Theory and Methods, the A.L.P.O. Saturn Handbook*, J. L. Benton, Jr., Associates in Astronomy 1995, 7th revised edition, p. 64

88. *The Giant Planet Jupiter*, J. H. Rogers, Cambridge University Press 1995, p. 59

89. *The Giant Planet Jupiter*, J. H. Rogers, Cambridge University Press 1995, p. 70

90. *The Giant Planet Jupiter*, J. H. Rogers, Cambridge University Press 1995, p. 70

91. *The Giant Planet Jupiter*, J. H. Rogers, Cambridge University Press 1995, p. 70

92. *The Giant Planet Jupiter*, J. H. Rogers, Cambridge University Press 1995, p. 70

93. *The Giant Planet Jupiter*, J. H. Rogers, Cambridge University Press 1995, p. 70

94. *The Giant Planet Jupiter*, J. H. Rogers, Cambridge University Press 1995, p. 289

95. "An HST Study of Jovian Chromophores", A. A. Simon-Miller, D. Banfield, and P. J. Gierasch. *Icarus* 2001, 149, p. 94

96. *The Giant Planet Jupiter*, J. H. Rogers, Cambridge University Press 1995, p. 280

97. *The Giant Planet Jupiter*, J. H. Rogers, Cambridge University Press 1995, p. 280

98. "An HST Study of Jovian Chromophores", A. A. Simon-Miller, D. Banfield, and P. J. Gierasch. *Icarus* 2001, 149, p. 94

99. *The Giant Planet Jupiter*, J. H. Rogers, Cambridge University Press 1995, p. 70

100. "Color and the Vertical Structure in Jupiter's Belts, Zones, and Weather Systems", A. A. Simon-Miller, D. Banfield, and P. J. Gierasch. *Icarus* 2001, 154, p. 459

101. *The Giant Planet Jupiter*, J. H. Rogers, Cambridge University Press 1995, p. 78

102. *Saturn and How to Observe It*, J. L. Benton, Jr., Springer 2005, p. 16

103. "Color and the Vertical Structure in Jupiter's Belts, Zones, and Weather Systems", A. A. Simon-Miller, D. Banfield, and P. J. Gierasch. *Icarus* 2001, 154, p. 459

104. "Color and the Vertical Structure in Jupiter's Belts, Zones, and Weather Systems", A. A. Simon-Miller, D. Banfield, and P. J. Gierasch. *Icarus* 2001, 154, p. 459

105. "Color and the Vertical Structure in Jupiter's Belts, Zones, and Weather Systems", A. A. Simon-Miller, D. Banfield, and P. J. Gierasch. *Icarus* 2001, 154, p. 464

106. "Color and the Vertical Structure in Jupiter's Belts, Zones, and Weather Systems", A. A. Simon-Miller, D. Banfield, and P. J. Gierasch. *Icarus* 2001, 154, p. 464

107. "Color and the Vertical Structure in Jupiter's Belts, Zones, and Weather Systems", A. A. Simon-Miller, D. Banfield, and P. J. Gierasch. *Icarus* 2001, 154, p. 466

108. "Color and the Vertical Structure in Jupiter's Belts, Zones, and Weather Systems", A. A. Simon-Miller, D. Banfield, and P. J. Gierasch. *Icarus* 2001, 154, p. 466

109. "Color and the Vertical Structure in Jupiter's Belts, Zones, and Weather Systems", A. A. Simon-Miller, D. Banfield, and P. J. Gierasch. *Icarus* 2001, 154, p. 467

110. "Color and the Vertical Structure in Jupiter's Belts, Zones, and Weather Systems", A. A. Simon-Miller, D. Banfield, and P. J. Gierasch. *Icarus* 2001, 154, p. 467

111. "Color and the Vertical Structure in Jupiter's Belts, Zones, and Weather Systems", A. A. Simon-Miller, D. Banfield, and P. J. Gierasch. *Icarus* 2001, 154, p. 467

112. "Color and the Vertical Structure in Jupiter's Belts, Zones, and Weather Systems", A. A. Simon-Miller, D. Banfield, and P. J. Gierasch. *Icarus* 2001, 154, p. 467

113. "Color and the Vertical Structure in Jupiter's Belts, Zones, and Weather Systems", A. A. Simon-Miller, D. Banfield, and P. J. Gierasch. *Icarus* 2001, 154, pp. 467, 468

114. *The Giant Planet Jupiter*, J. H. Rogers, Cambridge University Press 1995, p. 68

115. "Color and the Vertical Structure in Jupiter's Belts, Zones, and Weather Systems", A. A. Simon-Miller, D. Banfield, and P. J. Gierasch. *Icarus* 2001, 154, p. 469

116. *Jupiter Odyssey*, D. M. Harland, Springer/Praxis 2000, p. 259

117. *Jupiter Odyssey*, D. M. Harland, Springer/Praxis 2000, p. 259

118. *Jupiter Odyssey*, D. M. Harland, Springer/Praxis 2000, p. 258

119. "Color and the Vertical Structure in Jupiter's Belts, Zones, and Weather Systems", A. A. Simon-Miller, D. Banfield, and P. J. Gierasch. *Icarus* 2001, 154, p. 469

120. *Jupiter Odyssey*, D. M. Harland, Springer/Praxis 2000, pp. 259–260

121. *Jupiter Odyssey*, D. M. Harland, Springer/Praxis 2000, p. 260

122. *Jupiter Odyssey*, D. M. Harland, Springer/Praxis 2000, p. 260

123. "Characteristics of the Galileo Probe Entry Site from Earth-Based Remote Sensing Observations", G. S. Orton et al. *Journal of Geophysical Research* 1998, 103, p. 22,792

124. "Characteristics of the Galileo Probe Entry Site from Earth-Based Remote Sensing Observations", G. S. Orton et al. *Journal of Geophysical Research* 1998, 103, p. 22,808

125. "Characteristics of the Galileo Probe Entry Site from Earth-Based Remote Sensing Observations", G. S. Orton et al. *Journal of Geophysical Research* 1998, 103, p. 22,808

126. "Characteristics of the Galileo Probe Entry Site from Earth-Based Remote Sensing Observations", G. S. Orton et al. *Journal of Geophysical Research* 1998, 103, p. 22,812

127. "Characteristics of the Galileo Probe Entry Site from Earth-Based Remote Sensing Observations", Glenn S. Orton et al. *Journal of Geophysical Research* 1998, 103, p. 22,792

128. "Characteristics of the Galileo Probe entry Site from Earth-Based Remote Sensing Observations", G. S. Orton et al. *Journal of Geophysical Research* 1998, 103, p. 23,051

129. "Evolution and Persistence of 5-μm Hot Spots at the Galileo Probe Entry Latitude", J. L. Ortiz, G. S. Orton, A. J. Friedson, S. T. Stewart, B. M. Fisher, and J. R. Spencer. *Journal of Geophysical Research* 1998, 103, p. 23,063

130. "Evolution and Persistence of 5-μm Hot Spots at the Galileo Probe Entry Latitude", J. L. Ortiz, G. S. Orton, A. J. Friedson, S. T. Stewart, B. M. Fisher, and J. R. Spencer. *Journal of Geophysical Research* 1998, 103, p. 23,051

131. "Evolution and Persistence of 5-μm Hot Spots at the Galileo Probe Entry Latitude", J. L. Ortiz, G. S. Orton, A. J. Friedson, S. T. Stewart, B. M. Fisher, and J. R. Spencer. *Journal of Geophysical Research* 1998, 103, p. 23,058

132. "Evolution and Persistence of 5-μm Hot Spots at the Galileo Probe Entry Latitude", J. L. Ortiz, G. S. Orton, A. J. Friedson, S. T. Stewart, B. M. Fisher, and J. R. Spencer. *Journal of Geophysical Research* 1998, 103, p. 23,059

133. "Evolution and Persistence of 5-μm Hot Spots at the Galileo Probe Entry Latitude", J. L. Ortiz, G. S. Orton, A. J. Friedson, S. T. Stewart, B. M. Fisher, and J. R. Spencer. *Journal of Geophysical Research* 1998, 103, p. 23,060

134. *Jupiter Odyssey*, D. M. Harland, Springer/Praxis 2000, p. 278

135. *Jupiter Odyssey*, D. M. Harland, Springer/Praxis 2000, p. 279

136. "Jupiter's White Oval Turns Red", A. A. Simon-Miller, N. J. Chanover, G. S. Orton, M. Sussman, I. G. Tsavaris, and E. Karkoschka. *Icarus* 2006, 185, p. 558

137. "Jupiter's White Oval Turns Red", A. A. Simon-Miller, N. J. Chanover, G. S. Orton, M. Sussman, I. G. Tsavaris, and E. Karkoschka. *Icarus* 2006, 185, p. 559

138. "Jupiter's White Oval Turns Red", A. A. Simon-Miller, N. J. Chanover, G. S. Orton, M. Sussman, I. G. Tsavaris, and E. Karkoschka. *Icarus* 2006, 185, p. 559

139. "Jupiter's White Oval Turns Red", A. A. Simon-Miller, N. J. Chanover, G. S. Orton, M. Sussman, I. G. Tsavaris, and E. Karkoschka. *Icarus* 2006, 185, p. 560

140. *The Giant Planet Jupiter*, J. H. Rogers, Cambridge University Press 1995, p. 59

141. "An HST Study of Jovian Chromophores", A. A. Simon-Miller, D. Banfield, and P. J. Gierasch. *Icarus* 2001, 149, p. 104

142. *The Giant Planet Jupiter*, J. H. Rogers, Cambridge University Press 1995, p. 277

143. *The Giant Planet Jupiter*, J. H. Rogers, Cambridge University Press 1995, p. 277

144. *The Giant Planet Jupiter*, J. H. Rogers, Cambridge University Press 1995, p. 276

145. *The Giant Planet Jupiter*, J. H. Rogers, Cambridge University Press 1995, p. 277

146. *The Giant Planet Jupiter*, J. H. Rogers, Cambridge University Press 1995, p. 277

147. *The Giant Planet Jupiter*, J. H. Rogers, Cambridge University Press 1995, p. 277

148. *The Giant Planet Jupiter*, J. H. Rogers, Cambridge University Press 1995, p. 278

149. *The Giant Planet Jupiter*, J. H. Rogers, Cambridge University Press 1995, p. 279

150. *The Giant Planet Jupiter*, J. H. Rogers, Cambridge University Press 1995, p. 279

151. *The Giant Planet Jupiter*, J. H. Rogers, Cambridge University Press 1995, p. 279

152. *The Giant Planet Jupiter*, J. H. Rogers, Cambridge University Press 1995, p. 279

153. *The Giant Planet Jupiter*, J. H. Rogers, Cambridge University Press 1995, p. 279

154. *The Giant Planet Jupiter*, J. H. Rogers, Cambridge University Press 1995, p. 279

155. *The Giant Planet Jupiter*, J. H. Rogers, Cambridge University Press 1995, p. 279

156. "Jupiter's Atmospheric Composition from the Cassini Thermal Infrared Spectrometer Experiment", V. G. Kunde et al. *Science* 2004, 305, p. 1,582

157. "Jupiter's Atmospheric Composition from the Cassini Thermal Infrared Spectrometer Experiment", V. G. Kunde et al. *Science* 2004, 305, p. 1,582

158. "Jupiter's Atmospheric Composition from the Cassini Thermal Infrared Spectrometer Experiment", V. G. Kunde et al. *Science* 2004, 305, pp. 1,582, 1,585

159. "Jupiter's Atmospheric Composition from the Cassini Thermal Infrared Spectrometer Experiment", V. G. Kunde et al. *Science* 2004, 305, pp. 1,583, 1,584

160. *Jupiter Odyssey*, David M. Harland, Springer/Praxis 2000, pp. 115, 116

161. "Jupiter's Atmospheric Composition from the Cassini Thermal Infrared Spectrometer Experiment", V. G. Kunde et al. *Science* 2004, 305, (supporting on-line information, pp. 12–13.)

162. *Jupiter Odyssey*, D. M. Harland, Springer/Praxis 2000, p. 116

163. *The Giant Planet Jupiter*, J. H. Rogers, Cambridge University Press 1995, p. 279
164. *Jupiter Odyssey*, D. M. Harland, Springer/Praxis 2000, p. 249
165. *The Giant Planet Jupiter*, J. H. Rogers, Cambridge University Press 1995, p. 280
166. *Jupiter Odyssey*, D. M. Harland, Springer/Praxis 2000, p. 249
167. *The Giant Planet Jupiter*, J. H. Rogers, Cambridge University Press 1995, p. 280
168. *The Giant Planet Jupiter*, J. H. Rogers, Cambridge University Press 1995, p. 280
169. *Jupiter Odyssey*, D. M. Harland, Springer/Praxis 2000, p. 249
170. *The Giant Planet Jupiter*, J. H. Rogers, Cambridge University Press 1995, p. 282
171. *The Giant Planet Jupiter*, J. H. Rogers, Cambridge University Press 1995, p. 282
172. *The Giant Planet Jupiter*, J. H. Rogers, Cambridge University Press 1995, p. 282
173. *The Giant Planet Jupiter*, J. H. Rogers, Cambridge University Press 1995, p. 282
174. *The Giant Planet Jupiter*, J. H. Rogers, Cambridge University Press 1995, p. 66
175. *The Giant Planet Jupiter*, J. H. Rogers, Cambridge University Press 1995, p. 283
176. *Jupiter Odyssey*, D. M. Harland, Springer/Praxis 2000, p. 97
177. "Magnetic Moments at Jupiter", T. W. Hill. *Nature* 2002, 415, p. 965
178. *The Giant Planet Jupiter*, J. H. Rogers, Cambridge University Press 1995, p. 292
179. *The Giant Planet Jupiter*, J. H. Rogers, Cambridge University Press 1995, p. 292
180. *The Giant Planet Jupiter*, J. H. Rogers, Cambridge University Press 1995, p. 294
181. *The Giant Planet Jupiter*, J. H. Rogers, Cambridge University Press 1995, p. 294
182. "Magnetic Moments at Jupiter", T. W. Hill. *Nature* 2002, 415, p. 965
183. *Jupiter Odyssey*, D. M. Harland, Springer/Praxis 2000, p. 96
184. "The Dusk Flank of Jupiter's Magnetosphere", W. S. Kurth et al. *Nature* 2002, 415, p. 991
185. "The Dusk Flank of Jupiter's Magnetosphere", W. S. Kurth et al. *Nature* 2002, 415, p. 993
186. "The Dusk Flank of Jupiter's Magnetosphere", W. S. Kurth et al. *Nature* 2002, 415, p. 993
187. "The Dusk Flank of Jupiter's Magnetosphere", W. S. Kurth et al. *Nature* 2002, 415, pp. 992, 993
188. "The Dusk Flank of Jupiter's Magnetosphere", W. S. Kurth et al. *Nature* 2002, 415, pp. 992
189. "A Nebula of Gases from Io Surrounding Jupiter", S. M. Krinigis et al. *Nature* 2002, 415, p. 994
190. *Jupiter Odyssey*, D. M. Harland, Springer/Praxis 2000, pp. 408, 409
191. *Jupiter Odyssey*, D. M. Harland, Springer/Praxis 2000, pp. 36, 37
192. *The Giant Planet Jupiter*, J. H. Rogers, Cambridge University Press 1995, p. 294
193. *The Giant Planet Jupiter*, J. H. Rogers, Cambridge University Press 1995, p. 294
194. *The Giant Planet Jupiter*, J. H. Rogers, Cambridge University Press 1995, p. 294
195. *The Giant Planet Jupiter*, J. H. Rogers, Cambridge University Press 1995, p. 293
196. *The Giant Planet Jupiter*, J. H. Rogers, Cambridge University Press 1995, p. 298
197. *The Giant Planet Jupiter*, J. H. Rogers, Cambridge University Press 1995, p. 294
198. *The Giant Planet Jupiter*, J. H. Rogers, Cambridge University Press 1995, pp. 298, 303
199. *Neptune: the Planet, Rings, and Satellites*, E. D. Miner and R. R. Wessen, Springer/Praxis 2002, pp. 111–114
200. "The Dusk Flank of Jupiter's Magnetosphere", W. S. Kurth et al. *Nature* 2002, 415, pp. 992
201. "The Cassini-Huygens Flyby of Jupiter", C. J. Hansen, S. J. Bolton, D. L. Matson, L. J. Spilker, and J. -P. Lebreton. *Icarus* 2004, 172, p. 1
202. "The Cassini-Huygens Flyby of Jupiter", C. J. Hansen, S. J. Bolton, D. L. Matson, L. J. Spilker, and J. -P. Lebreton. *Icarus* 2004, 172, p. 4
203. "Magnetic Moments at Jupiter", T. W. Hill. *Nature* 2002, 415, p. 966
204. *The Giant Planet Jupiter*, J. H. Rogers, Cambridge University Press 1995, p. 303
205. *Jupiter Odyssey*, D. M. Harland, Springer/Praxis 2000, p. 270
206. *The Giant Planet Jupiter*, J. H. Rogers, Cambridge University Press 1995, p. 302
207. *Jupiter Odyssey*, D. M. Harland, Springer/Praxis 2000, p. 271
208. *The Giant Planet Jupiter*, J. H. Rogers, Cambridge University Press 1995, p. 302
209. *The Giant Planet Jupiter*, J. H. Rogers, Cambridge University Press 1995, p. 295
210. *The Giant Planet Jupiter*, J. H. Rogers, Cambridge University Press 1995, p. 302
211. *The Giant Planet Jupiter*, J. H. Rogers, Cambridge University Press 1995, p. 303
212. *Jupiter Odyssey*, D. M. Harland, Springer/Praxis 2000, pp. 101, 102

213. *The Giant Planet Jupiter*, J. H. Rogers, Cambridge University Press 1995, p. 303
214. *The Giant Planet Jupiter*, J. H. Rogers, Cambridge University Press 1995, p. 303
215. *Jupiter Odyssey*, D. M. Harland, Springer/Praxis 2000, p. 322
216. *The Giant Planet Jupiter*, J. H. Rogers, Cambridge University Press 1995, p. 303
217. *The Giant Planet Jupiter*, J. H. Rogers, Cambridge University Press 1995, p. 303
218. *The Giant Planet Jupiter*, J. H. Rogers, Cambridge University Press 1995, p. 303
219. *Jupiter Odyssey*, D. M. Harland, Springer/Praxis 2000, p. 101
220. *The Giant Planet Jupiter*, J. H. Rogers, Cambridge University Press 1995, p. 303
221. *The Giant Planet Jupiter*, J. H. Rogers, Cambridge University Press 1995, p. 303
222. *The Giant Planet Jupiter*, J. H. Rogers, Cambridge University Press 1995, p. 305
223. *The Giant Planet Jupiter*, J. H. Rogers, Cambridge University Press 1995, p. 306
224. *The Giant Planet Jupiter*, J. H. Rogers, Cambridge University Press 1995, pp. 306, 307
225. *Jupiter Odyssey*, D. M. Harland, Springer/Praxis 2000, p. 101
226. *Jupiter Odyssey*, D. M. Harland, Springer/Praxis 2000, p. 332
227. *Jupiter Odyssey*, D. M. Harland, Springer/Praxis 2000, pp. 101, 102
228. *Jupiter Odyssey*, D. M. Harland, Springer/Praxis 2000, pp. 112, 113
229. *The Giant Planet Jupiter*, J. H. Rogers, Cambridge University Press 1995, p. 288
230. *Jupiter Odyssey*, D. M. Harland, Springer/Praxis 2000, p. 257
231. *The Giant Planet Jupiter*, J. H. Rogers, Cambridge University Press 1995, p. 288
232. *The Giant Planet Jupiter*, J. H. Rogers, Cambridge University Press 1995, p. 288
233. *The Giant Planet Jupiter*, J. H. Rogers, Cambridge University Press 1995, pp. 288–289
234. *The Giant Planet Jupiter*, J. H. Rogers, Cambridge University Press 1995, p. 288
235. *Jupiter Odyssey*, D. M. Harland, Springer/Praxis 2000, p. 257
236. *Jupiter Odyssey*, D. M. Harland, Springer/Praxis 2000, p. 272
237. "An Auroral Flare at Jupiter", J. H. Waite, Jr. et al. *Nature* 2001, 410, p. 787
238. "An Auroral Flare at Jupiter", J. H. Waite, Jr. et al. *Nature* 2001, 410, p. 787
239. "An Auroral Flare at Jupiter", J. H. Waite, Jr. et al. *Nature* 2001, 410, p. 787
240. "An Auroral Flare at Jupiter", J. H. Waite, Jr. et al. *Nature* 2001, 410, p. 788
241. "An Auroral Flare at Jupiter", J. H. Waite, Jr. et al. *Nature* 2001, 410, pp. 788–789
242. "An Auroral Flare at Jupiter", J. H. Waite, Jr. et al. *Nature* 2001, 410, p. 787
243. "The Cassini-Huygens Flyby of Jupiter", C. J. Hansen, S. J. Bolton, D. L. Matson, L. J. Spilker, and J.-P. Lebreton. *Icarus* 2004, 172, p.4
244. "Magnetic Moments at Jupiter", T. W. Hill. *Nature* 2002, 415, p. 966
245. "Auroral Structures at Jupiter and Earth", T. W. Hill. *Advances in Space Research* 2004, 33, pp. 2021, 2025
246. "The Cassini-Huygens Flyby of Jupiter", C. J. Hansen, S. J. Bolton, D. L. Matson, L. J. Spilker, and J.-P. Lebreton. *Icarus* 2004, 172, p.4
247. "Transient Aurora on Jupiter from Injections of Magnetospheric Electrons", J. Mauk, J. T. Clarke, D. Grodent, J. H. Waite, Jr., C. P. Paranicas, and D. J. Williams. *Nature* 2002, 415, p. 1003
248. *Jupiter Odyssey*, D. M. Harland, Springer/Praxis 2000, p. 272
249. "Magnetic Moments at Jupiter", T. W. Hill. *Nature* 2002, 415, p. 966
250. "Auroral Structures at Jupiter and Earth", T. W. Hill. *Advances in Space Research* 2004, 33, p. 2030
251. "Control of Jupiter's Radio Emission and Aurorae by the Solar Wind", D. A. Gurnet et al. *Nature* 2002, 415, p. 985
252. *The Giant Planet Jupiter*, J. H. Rogers, Cambridge University Press 1995, p. 298
253. *The Giant Planet Jupiter*, J. H. Rogers, Cambridge University Press 1995, p. 299
254. *The Giant Planet Jupiter*, J. H. Rogers, Cambridge University Press 1995, p. 298
255. *The Giant Planet Jupiter*, J. H. Rogers, Cambridge University Press 1995, p. 299
256. *The Giant Planet Jupiter*, J. H. Rogers, Cambridge University Press 1995, p. 299
257. "Control of Jupiter's Radio Emission and Aurorae by the Solar Wind", D. A. Gurnet et al. *Nature* 2002, 415, p. 985
258. "Control of Jupiter's Radio Emission and Aurorae by the Solar Wind", D. A. Gurnet et al. *Nature* 2002, 415, p. 985

259. "Control of Jupiter's Radio Emission and Aurorae by the Solar Wind", D. A. Gurnet et al. *Nature* 2002, 415, p. 989
260. "Control of Jupiter's Radio Emission and Aurorae by the Solar Wind", D. A. Gurnet et al. *Nature* 2002, 415, pp. 986–988
261. "Control of Jupiter's Radio Emission and Aurorae by the Solar Wind", D. A. Gurnet et al. *Nature* 2002, 415, p. 986
262. *The Giant Planet Jupiter*, J. H. Rogers, Cambridge University Press 1995, p. 287
263. *Jupiter Odyssey*, D. M. Harland, Springer/Praxis 2000, p. 124
264. "Lightning on Jupiter Observed in the H_α Line by the Cassini Imaging Science Subsystem", U. A. Dyudina et al. *Icarus* 2004, 172, p. 24
265. *Jupiter Odyssey*, D. M. Harland, Springer/Praxis 2000, p. 124
266. *Jupiter Odyssey*, D. M. Harland, Springer/Praxis 2000, p. 124
267. *Jupiter Odyssey*, D. M. Harland, Springer/Praxis 2000, pp. 257, 258
268. "Galileo Images of Lightning on Jupiter", B. Little et al. *Icarus* 1999, 142, p. 318
269. *Jupiter Odyssey*, D. M. Harland, Springer/Praxis 2000, p. 273
270. "Lightning on Jupiter Observed in the H_α Line by the Cassini Imaging Science Subsystem", U. A. Dyudina et al. *Icarus* 2004, 172, p. 25
271. "Lightning on Jupiter Observed in the H_α Line by the Cassini Imaging Science Subsystem", U. A. Dyudina et al. *Icarus* 2004, 172, p. 32
272. "Galileo Images of Lightning on Jupiter", B. Little et al. *Icarus* 1999, 142, pp. 313, 314
273. "Lightning on Jupiter Observed in the H_α Line by the Cassini Imaging Science Subsystem", U. A. Dyudina et al. *Icarus* 2004, 172, p. 32
274. "Galileo Images of Lightning on Jupiter", B. Little et al. *Icarus* 1999, 142, p. 306
275. "Lightning on Jupiter Observed in the H_α Line by the Cassini Imaging Science Subsystem", U. A. Dyudina et al. *Icarus* 2004, 172, p. 32
276. "Lightning on Jupiter Observed in the H_α Line by the Cassini Imaging Science Subsystem", U. A. Dyudina et al. *Icarus* 2004, 172, p. 32
277. "Lightning on Jupiter Observed in the H_α Line by the Cassini Imaging Science Subsystem", U. A. Dyudina et al. *Icarus* 2004, 172, pp. 33, 34
278. *Jupiter Odyssey*, D. M. Harland, Springer/Praxis 2000, p. 298
279. *The Giant Planet Jupiter*, J. H. Rogers, Cambridge University Press 1995, p. 307
280. "Ultraviolet emissions from the Magnetic Footprints of Io, Ganymede, and Europa on Jupiter", J. T. Clarke et al. *Nature* 2002, 415, p. 997
281. "Ultraviolet emissions from the Magnetic Footprints of Io, Ganymede, and Europa on Jupiter", J. T. Clarke et al. *Nature* 2002, 415, p. 997
282. "Auroral Structures at Jupiter and Earth", T. W. Hill. *Advances in Space Research* 2004, 33, p. 2030
283. "Ultraviolet emissions from the Magnetic Footprints of Io, Ganymede, and Europa on Jupiter", J. T. Clarke et al. *Nature* 2002, 415, p. 997
284. *Jupiter Odyssey*, D. M. Harland, Springer/Praxis 2000, p. 273
285. *Jupiter Odyssey*, D. M. Harland, Springer/Praxis 2000, p. 298
286. "Ultraviolet emissions from the Magnetic Footprints of Io, Ganymede, and Europa on Jupiter", J. T. Clarke et al. *Nature* 2002, 415, p. 999
287. "Simultaneous Chandra X ray, Hubble Space Telescope Ultraviolet, and Ulysses Radio Observations of Jupiter's Aurora", R. F. Elsner et al. *Journal of Geophysical Research* 2005, 110, p. 417
288. "Simultaneous Chandra X ray, Hubble Space Telescope Ultraviolet, and Ulysses Radio Observations of Jupiter's Aurora", R. F. Elsner et al. *Journal of Geophysical Research* 2005, 110, p. 418
289. "Simultaneous Chandra X ray, Hubble Space Telescope Ultraviolet, and Ulysses Radio Observations of Jupiter's Aurora", R. F. Elsner et al. *Journal of Geophysical Research* 2005, 110, p. 418
290. "Simultaneous Chandra X ray, Hubble Space Telescope Ultraviolet, and Ulysses Radio Observations of Jupiter's Aurora", R. F. Elsner et al. *Journal of Geophysical Research* 2005, 110, p. 419

291. "A Pulsating Auroral X-ray Hot Spot on Jupiter", G. R. Gladstone et al. *Nature* 2002, 415, p. 1000

292. "Simultaneous Chandra X ray, Hubble Space Telescope Ultraviolet, and Ulysses Radio Observations of Jupiter's Aurora", R. F. Elsner et al. *Journal of Geophysical Research* 2005, 110, p. 419

293. "Simultaneous Chandra X ray, Hubble Space Telescope Ultraviolet, and Ulysses Radio Observations of Jupiter's Aurora", R. F. Elsner et al. *Journal of Geophysical Research* 2005, 110, p. 419

294. "Simultaneous Chandra X ray, Hubble Space Telescope Ultraviolet, and Ulysses Radio Observations of Jupiter's Aurora", R. F. Elsner et al. *Journal of Geophysical Research* 2005, 110, pp. 419–420

295. "Simultaneous Chandra X ray, Hubble Space Telescope Ultraviolet, and Ulysses Radio Observations of Jupiter's Aurora", R. F. Elsner et al. *Journal of Geophysical Research* 2005, 110, p. 420

296. "A Pulsating Auroral X-ray Hot Spot on Jupiter", G. R. Gladstone et al. *Nature* 2002, 415, p. 1,000

297. "Simultaneous Chandra X ray, Hubble Space Telescope Ultraviolet, and Ulysses Radio Observations of Jupiter's Aurora", R. F. Elsner et al. *Journal of Geophysical Research* 2005, 110, p. 421

298. "Simultaneous Chandra X ray, Hubble Space Telescope Ultraviolet, and Ulysses Radio Observations of Jupiter's Aurora", R. F. Elsner et al. *Journal of Geophysical Research* 2005, 110, p. 422

299. "Simultaneous Chandra X ray, Hubble Space Telescope Ultraviolet, and Ulysses Radio Observations of Jupiter's Aurora", R. F. Elsner et al. *Journal of Geophysical Research* 2005, 110, p. 426

300. *The Giant Planet Jupiter*, J. H. Rogers, Cambridge University Press 1995, pp. 334–336

301. "Surface Changes on Io During the Galileo Mission", P. Geissler, A. McEwen, C. Phillips, L. Keszthelyi, and J. Spencer. *Icarus* 2004, 169, p. 30

302. "Surface Changes on Io During the Galileo Mission", P. Geissler, A. McEwen, C. Phillips, L. Keszthelyi, and J. Spencer. *Icarus* 2004, 169, p. 29

303. "Core Sizes and Internal Structure of Earth's and Jupiter's Satellites", O. L. Kuskov and V. A. Kronrod. *Icarus* 2001, 151, p. 221

304. "Implications from Galileo Observations on the Interior Structure and Chemistry of the Galilean Satellites", F. Sohl, T. Spohn, D. Breuer, and K. Nagel. *Icarus* 2002, 157, p. 105

305. "Core Sizes and Internal Structure of Earth's and Jupiter's Satellites", O. L. Kuskov and V. A. Kronrod. *Icarus* 2001, 151, p. 221

306. *The Giant Planet Jupiter*, J. H. Rogers, Cambridge University Press 1995, p. 337

307. "Surface Changes on Io During the Galileo Mission", P. Geissler, A. McEwen, C. Phillips, L. Keszthelyi, and J. Spencer. *Icarus* 2004, 169, pp. 30, 32–34

308. *Jupiter Odyssey*, D. M. Harland, Springer/Praxis 2000, p. 291

309. *The Giant Planet Jupiter*, J. H. Rogers, Cambridge University Press 1995, pp. 357, 358

310. "Surface Changes on Io During the Galileo Mission", P. Geissler, A. McEwen, C. Phillips, L. Keszthelyi, and J. Spencer. *Icarus* 2004, 169, p. 29

311. "Surface Changes on Io During the Galileo Mission", P. Geissler, A. McEwen, C. Phillips, L. Keszthelyi, and J. Spencer. *Icarus* 2004, 169, p. 59

312. *Jupiter Odyssey*, D. M. Harland, Springer/Praxis 2000, pp. 297–298

313. "Surface Changes on Io During the Galileo Mission", P. Geissler, A. McEwen, C. Phillips, L. Keszthelyi, and J. Spencer. *Icarus* 2004, 169, p. 29

314. "The Final Galileo SSI Observations of Io: Orbits G28 – I33", E. P. Turtle et al. *Icarus* 2004, 169, p. 6

315. "The Final Galileo SSI Observations of Io: Orbits G28 – I33", E. P. Turtle et al. *Icarus* 2004, 169, p. 8

316. "Surface Changes on Io During the Galileo Mission", P. Geissler, A. McEwen, C. Phillips, L. Keszthelyi, and J. Spencer. *Icarus* 2004, 169, p. 31

317. "Surface Changes on Io During the Galileo Mission", P. Geissler, A. McEwen, C. Phillips, L. Keszthelyi, and J. Spencer. *Icarus* 2004, 169, pp. 36, 37

318. "Surface Changes on Io During the Galileo Mission", Paul Geissler, Alfred McEwen, Cynthia Phillips, Laszlo Keszthelyi, and John Spencer. *Icarus* 2004, 169, p. 44

319. "Surface Changes on Io During the Galileo Mission", P. Geissler, A. McEwen, C. Phillips, L. Keszthelyi, and J. Spencer. *Icarus* 2004, 169, p. 51

320. "Surface Changes on Io During the Galileo Mission", P. Geissler, A. McEwen, C. Phillips, L. Keszthelyi, and J. Spencer. *Icarus* 2004, 169, p. 59

321. "Surface Changes on Io During the Galileo Mission", P. Geissler, A. McEwen, C. Phillips, L. Keszthelyi, and J. Spencer. *Icarus* 2004, 169, p. 31

322. *Jupiter Odyssey*, David M. Harland, Springer/Praxis 2000, pp. 314, 315

323. "Surface Changes on Io During the Galileo Mission", P. Geissler, A. McEwen, C. Phillips, L. Keszthelyi, and J. Spencer. *Icarus* 2004, 169, p. 58

324. "Surface Changes on Io During the Galileo Mission", P. Geissler, A. McEwen, C. Phillips, L. Keszthelyi, and J Spencer. *Icarus* 2004, 169, pp. 29, 61

325. "Surface Changes on Io During the Galileo Mission", P. Geissler, A. McEwen, C. Phillips, L. Keszthelyi, and J. Spencer. *Icarus* 2004, 169, p. 62

326. "Surface Changes on Io During the Galileo Mission", P. Geissler, A. McEwen, C. Phillips, L. Keszthelyi, and J. Spencer. *Icarus* 2004, 169, p. 62

327. "Surface Changes on Io During the Galileo Mission", P. Geissler, A. McEwen, C. Phillips, L. Keszthelyi, and J. Spencer. *Icarus* 2004, 169, pp. 56, 57

328. "Surface Changes on Io During the Galileo Mission", P. Geissler, A. McEwen, C. Phillips, L. Keszthelyi, and J. Spencer. *Icarus* 2004, 169, p. 62

329. "The Final Galileo SSI Observations of Io: Orbits G28 – I33", E. P. Turtle et al. *Icarus* 2004, 169, pp. 3, 24

330. *Jupiter Odyssey*, D. M. Harland, Springer/Praxis 2000, p. 341

331. *Jupiter Odyssey*, D. M. Harland, Springer/Praxis 2000, p. 349

332. "The Final Galileo SSI Observations of Io: Orbits G28 – I33", E. P. Turtle et al. *Icarus* 2004, 169, p. 24

333. "Surface Changes on Io During the Galileo Mission", P. Geissler, A. McEwen, C. Phillips, L. Keszthelyi, and J. Spencer. *Icarus* 2004, 169, p. 58

334. "Surface Changes on Io During the Galileo Mission", P. Geissler, A. McEwen, C. Phillips, L. Keszthelyi, and J. Spencer. *Icarus* 2004, 169, p. 37

335. "Surface Changes on Io During the Galileo Mission", P. Geissler, A. McEwen, C. Phillips, L. Keszthelyi, and J. Spencer. *Icarus* 2004, 169, p. 58

336. "Surface Changes on Io During the Galileo Mission", P. Geissler, A. McEwen, C. Phillips, L. Keszthelyi, and J. Spencer. *Icarus* 2004, 169, p. 53

337. "Surface Changes on Io During the Galileo Mission", P. Geissler, A. McEwen, C. Phillips, L. Keszthelyi, and J. Spencer. *Icarus* 2004, 169, p. 61

338. "Surface Changes on Io During the Galileo Mission", P. Geissler, A. McEwen, C. Phillips, L. Keszthelyi, and J. Spencer. *Icarus* 2004, 169, p. 41

339. "Surface Changes on Io During the Galileo Mission", P. Geissler, A. McEwen, C. Phillips, L. Keszthelyi, and J. Spencer. *Icarus* 2004, 169, p. 61

340. "The Final Galileo SSI Observations of Io: Orbits G28 – I33", E. P. Turtle et al. *Icarus* 2004, 169, p. 3

341. *Jupiter Odyssey*, D. M. Harland, Springer/Praxis 2000, p. 346

342. "Ridges and Tidal Stress on Io", G. D. Bart, E. P. Turtle, W. L. Jaeger, L. P. Keszthelyi, and R. Greenberg. *Icarus* 2004, 169, p. 123

343. "Implications from Galileo Observations on the Interior Structure and Chemistry of the Galilean Satellites", F. Sohl, T. Spohn, D. Breuer, and K. Nagel. *Icarus* 2002, 157, p. 104

344. "Surface Changes on Io During the Galileo Mission", P. Geissler, A. McEwen, C. Phillips, L. Keszthelyi, and J. Spencer. *Icarus* 2004, 169, p. 61

345. *Jupiter Odyssey*, D. M. Harland, Springer/Praxis 2000, p. 294

346. *Jupiter Odyssey*, D. M. Harland, Springer/Praxis 2000, pp. 345, 348

347. *Jupiter Odyssey*, D. M. Harland, Springer/Praxis 2000, pp. 347, 348

348. *Jupiter Odyssey*, D. M. Harland, Springer/Praxis 2000, p. 345

349. *Jupiter Odyssey*, D. M. Harland, Springer/Praxis 2000, pp. 347, 353

350. "Ridges and Tidal Stress on Io", G. D. Bart, E. P. Turtle, W. L. Jaeger, L. P. Keszthelyi, and R. Greenberg. *Icarus* 2004, 169, pp. 111, 113

351. "Ridges and Tidal Stress on Io", G. D. Bart, E. P. Turtle, W. L. Jaeger, L. P. Keszthelyi, and R. Greenberg. *Icarus* 2004, 169, p. 124

352. "Ridges and Tidal Stress on Io", G. D. Bart, E. P. Turtle, W. L. Jaeger, L. P. Keszthelyi, and R. Greenberg. *Icarus* 2004, 169, p. 125

353. "Core Sizes and Internal Structure of Earth's and Jupiter's Satellites", O. L. Kuskov and V. A. Kronrod. *Icarus* 2001, 151, p. 204

354. *Jupiter Odyssey*, D. M. Harland, Springer/Praxis 2000, p. 300

355. *Jupiter Odyssey*, D. M. Harland, Springer/Praxis 2000, pp. 297, 318, 319, 335

356. "Implications from Galileo Observations on the Interior Structure and Chemistry of the Galilean Satellites", F. Sohl, T. Spohn, D. Breuer, and K. Nagel. *Icarus* 2002, 157, p. 105

357. "Core Sizes and Internal Structure of Earth's and Jupiter's Satellites", O. L. Kuskov and V. A. Kronrod. *Icarus* 2001, 151, p. 221

358. *The Giant Planet Jupiter*, J. H. Rogers, Cambridge University Press 1995, p. 331

359. "The Chemical Nature of Europa Surface Material and the Relation to a Subsurface Ocean", T. M. Orlando, T. B. McCord, and G. A. Grieves. *Icarus* 2005, 177, p. 529

360. *Jupiter Odyssey*, D. M. Harland, Springer/Praxis 2000, p. 196

361. "Cassini UVIS Observations of Europa's Oxygen Atmosphere and Torus", C. J. Hansen, D. E. Shemansky, and A.R. Hendrix. *Icarus* 2005, 176, pp. 305, 306, 313

362. "Cassini UVIS Observations of Europa's Oxygen Atmosphere and Torus", C. J. Hansen, D. E. Shemansky, and A.R. Hendrix. *Icarus* 2005, 176, p. 305

363. *Jupiter Odyssey*, D. M. Harland, Springer/Praxis 2000, p. 183

364. "The Chemical Nature of Europa Surface Material and the Relation to a Subsurface Ocean", T. M. Orlando, T. B. McCord, and G. A. Grieves. *Icarus* 2005, 177, p. 532

365. "Effects of Plasticity on Convection in an Ice Shell: Implications for Europa", A. P. Showman and L. Han. *Icarus* 2005, 177, pp. 425, 426

366. "Resurfacing History of Europa from Pole-to-Pole Geological Mapping", P. H. Figueredo and R. Greeley. *Icarus* 2004, 167, p. 287

367. "Putative Ice Flows on Europa: Geometric Patterns and Relation to Topography Collectively Constrain Material Properties and Effusion Rates", H. Miyamoto, G. Mitri, A. P. Showman, and J. M. Dohm. *Icarus* 2005, 177, pp. 413, 414

368. *Jupiter Odyssey*, D. M. Harland, Springer/Praxis 2000, p. 183

369. *Jupiter Odyssey*, D. M. Harland, Springer/Praxis 2000, p. 185

370. "Resurfacing History of Europa from Pole-to-Pole Geological Mapping", P. H. Figueredo and R. Greeley. *Icarus* 2004, 167, p. 282

371. *Jupiter Odyssey*, D. M. Harland, Springer/Praxis 2000, p. 189

372. *Jupiter Odyssey*, D. M. Harland, Springer/Praxis 2000, pp. 198, 199

373. "Resurfacing History of Europa from Pole-to-Pole Geological Mapping", P. H. Figueredo and R. Greeley. *Icarus* 2004, 167, p. 292

374. *Jupiter Odyssey*, D. M. Harland, Springer/Praxis 2000, pp. 230–231, 232

375. "Mechanics of Tidally Driven Fractures in Europa's Ice Shell", S. Lee, R. T. Pappalardo, and N. C. Makris. *Icarus* 2005, 177, p. 368

376. "Resurfacing History of Europa from Pole-to-Pole Geological Mapping", P. H. Figueredo and R. Greeley. *Icarus* 2004, 167, p. 289

377. "The Temperature of Europa's Subsurface Water Ocean", H. J. Melosh, A. G. Ekholm, A. P. Showman, and R. D. Lorenz. *Icarus* 2004, 168, p. 500

378. "Chaos on Europa", R. Greenberg, G. V. Hoppa, B. R. Tufts, P. Geissler, and J. Riley. *Icarus* 1999, 141, pp. 263, 269

379. "Chaos on Europa", R. Greenberg, G. V. Hoppa, B. R. Tufts, P. Geissler, and J. Riley. *Icarus* 1999, 141, pp. 283, 284

380. "Resurfacing History of Europa from Pole-to-Pole Geological Mapping", P. H. Figueredo and R. Greeley. *Icarus* 2004, 167, p. 287

381. "Resurfacing History of Europa from Pole-to-Pole Geological Mapping", P. H. Figueredo and R. Greeley. *Icarus* 2004, 167, p. 287

382. *Jupiter Odyssey*, D. M. Harland, Springer/Praxis 2000, pp. 185, 192

383. "Impact Features on Europa: Results of the Galileo Europa Mission (GEM)", J. M. Moore et al. *Icarus* 2001, 151, p. 95

384. "Impact Features on Europa: Results of the Galileo Europa Mission (GEM)", J. M. Moore et al. *Icarus* 2001, 151, p. 97

385. "The Heat Flow of Europa", J. Ruiz. *Icarus* 2005, 177, p. 438

386. "Impact Features on Europa: Results of the Galileo Europa Mission (GEM)", J. M. Moore et al. *Icarus* 2001, 151, p. 109

387. "Impact Features on Europa: Results of the Galileo Europa Mission (GEM)", J. M. Moore et al. *Icarus* 2001, 151, pp. 98, 99

388. "Impact Features on Europa: Results of the Galileo Europa Mission (GEM)", Jeffrey M. Moore et al. *Icarus* 2001, 151, pp. 97, 98

389. "Impact Features on Europa: Results of the Galileo Europa Mission (GEM)", J. M. Moore et al. *Icarus* 2001, 151, p. 107

390. "Europa's Crust and Ocean: Origin, Composition, and the Prospects for Life", J. S. Kargel et al. *Icarus* 2000, 148, p. 226

391. "Europa's Icy Shell: Past and Present State, and Future Exploration", F. Nimmo et al. *Icarus* 2005, 177, p. 294

392. "Impact Features on Europa: Results of the Galileo Europa Mission (GEM)", J. M. Moore et al. *Icarus* 2001, 151, p. 95

393. "Impact Features on Europa: Results of the Galileo Europa Mission (GEM)", J. M. Moore et al. *Icarus* 2001, 151, p. 110

394. "Resurfacing History of Europa from Pole-to-Pole Geological Mapping", P. H. Figueredo and R. Greeley. *Icarus* 2004, 167, pp. 287, 306

395. "Resurfacing History of Europa from Pole-to-Pole Geological Mapping", P. H. Figueredo and R. Greeley. *Icarus* 2004, 167, pp. 299, 305

396. "Resurfacing History of Europa from Pole-to-Pole Geological Mapping", P. H. Figueredo and R. Greeley. *Icarus* 2004, 167, p. 305

397. "Europa's Crust and Ocean: Origin, Composition, and the Prospects for Life", J. S. Kargel et al. *Icarus* 2000, 148, p. 249

398. "Core Sizes and Internal Structure of Earth's and Jupiter's Satellites", O. L. Kuskov and V. A. Kronrod. *Icarus* 2001, 151, p. 204

399. "Core Sizes and Internal Structure of Earth's and Jupiter's Satellites", O. L. Kuskov and V. A. Kronrod. *Icarus* 2001, 151, p. 216

400. "Implications from Galileo Observations on the Interior Structure and Chemistry of the Galilean Satellites", F. Sohl, T. Spohn, D. Breuer, and K. Nagel. *Icarus* 2002, 157, pp. 104, 118

401. "Implications from Galileo Observations on the Interior Structure and Chemistry of the Galilean Satellites", F. Sohl, T. Spohn, D. Breuer, and K. Nagel. *Icarus* 2002, 157, pp. 104, 110, 118

402. "Core Sizes and Internal Structure of Earth's and Jupiter's Satellites", O. L. Kuskov and V. A. Kronrod. *Icarus* 2001, 151, p. 214

403. "Implications from Galileo Observations on the Interior Structure and Chemistry of the Galilean Satellites", F. Sohl, T. Spohn, D. Breuer, and K. Nagel. *Icarus* 2002, 157, p. 105

404. "Implications from Galileo Observations on the Interior Structure and Chemistry of the Galilean Satellites", F. Sohl, T. Spohn, D. Breuer, and K. Nagel. *Icarus* 2002, 157, p. 105

405. "Subsurface Oceans on Europa and Callisto: Constraints from Galileo Magnetometer Observations", C. Zimmer and K. K. Khurana. *Icarus* 2000, 147, p. 239

406. "Europa's Icy Shell: Past and Present State, and Future Exploration", F. Nimmo et al. *Icarus* 2005, 177, p. 294

407. "The Chemical Nature of Europa Surface Material and the Relation to a Subsurface Ocean", T. M. Orlando, T. B. McCord, and G. A. Grieves. *Icarus* 2005, 177, p. 529

408. "Europa's Crust and Ocean: Origin, Composition, and the Prospects for Life", J. S. Kargel et al. *Icarus* 2000, 148, p. 228

409. "The Great Thickness Debate: Ice shell Thickness Models for Europa and Comparisons with Estimates Based on Flexure at Ridges", S. E. Billings and S. A. Katterhorn. *Icarus* 2005, 177, p. 397

410. "The Temperature of Europa's Subsurface Water Ocean", H. J. Melosh, A. G. Ekholm, A. P. Showman, and R. D. Lorenz. *Icarus* 2004, 168, p. 498

411. "Europa's Crust and Ocean: Origin, Composition, and the Prospects for Life", J. S. Kargel et al. *Icarus* 2000, 148, pp. 251–253, 256

412. "Habitats and Taphonomy of Europa", J. H. Lipps and S. Rieboldt. *Icarus* 2005, 177, p. 515

413. *Jupiter Odyssey*, D. M. Harland, Springer/Praxis 2000, p. 242

414. "Internal Structure of Europa and Callisto", O. L. Kuskov and V. A. Kronrod. *Icarus* 2005, 177, p. 559

415. *The Giant Planet Jupiter*, J. H. Rogers, Cambridge University Press 1995, p. 331

416. "Core Sizes and Internal Structure of Earth's and Jupiter's Satellites", O. L. Kuskov and V. A. Kronrod. *Icarus* 2001, 151, p. 221

417. "Implications from Galileo Observations on the Interior Structure and Chemistry of the Galilean Satellites", F. Sohl, T. Spohn, D. Breuer, and K. Nagel. *Icarus* 2002, 157, p. 105

418. *Jupiter Odyssey*, D. M. Harland, Springer/Praxis 2000, p. 133

419. *Jupiter Odyssey*, D. M. Harland, Springer/Praxis 2000, p. 152

420. *Jupiter Odyssey*, D. M. Harland, Springer/Praxis 2000, p. 134

421. "Formation of Grooved Terrain on Ganymede: Extensional Instability Mediated by Cold, Superplastic Creep", A. J. Dombard and W. B. McKinnon. *Icarus* 2001, 154, p. 321

422. "On the Resurfacing of Ganymede by Liquid – Water Volcanism", A. P. Showman, I. Mosqueira, and J. W. Head III. *Icarus* 2004, 172, pp. 625, 626, 628

423. "Formation of Ganymede Grooved Terrain by Sequential Extensional Episodes: Implications of Galileo Observations for Regional Stratigraphy", G. C. Collins. J. W. Head, and R. T. Pappalardo. *Icarus* 1998, 135, p. 346

424. "Grooved Terrain on Ganymede: First Results from Galileo High-Resolution Imaging", R. T. Pappalardo et al. *Icarus* 1998, 135, pp. 276–278

425. "Grooved Terrain on Ganymede: First Results from Galileo High-Resolution Imaging", R. T. Pappalardo et al. *Icarus* 1998, 135, p. 300

426. "Dark Terrain on Ganymede: Geological Mapping and Interpretation of Galileo Regio at High Resolution", L. M. Prockter et al. *Icarus* 1998, 135, pp. 317, 318, 329, 338

427. "Dark Terrain on Ganymede: Geological Mapping and Interpretation of Galileo Regio at High Resolution", Louise M. Prockter et al. *Icarus* 1998, 135, p. 342

428. "Morphology and Origin of Palimpsests on Ganymede Based on Galileo Observations", K. B. Jones, J. W. Head III, R. T. Pappalardo, and J. M. Moore. *Icarus* 2003, 164, p. 197

429. *The Giant Planet Jupiter*, J. H. Rogers, Cambridge University Press 1995, p. 375

430. *Jupiter Odyssey*, D. M. Harland, Springer/Praxis 2000, p. 134

431. *Jupiter Odyssey*, D. M. Harland, Springer/Praxis 2000, pp. 142, 143

432. *Jupiter Odyssey*, D. M. Harland, Springer/Praxis 2000, p. 150

433. *Jupiter Odyssey*, D. M. Harland, Springer/Praxis 2000, p. 159

434. "On the Resurfacing of Ganymede by Liquid – Water Volcanism", A. P. Showman, I. Mosqueira, and J. W. Head III. *Icarus* 2004, 172, p. 626

435. "Core Sizes and Internal Structure of Earth's and Jupiter's Satellites", O. L. Kuskov and V. A. Kronrod. *Icarus* 2001, 151, pp. 204, 214

436. "Implications from Galileo Observations on the Interior Structure and Chemistry of the Galilean Satellites", F. Sohl, T. Spohn, D. Breuer, and K. Nagel. *Icarus* 2002, 157, p. 104

437. "Implications from Galileo Observations on the Interior Structure and Chemistry of the Galilean Satellites", F. Sohl, T. Spohn, D. Breuer, and K. Nagel. *Icarus* 2002, 157, pp. 104, 106

438. *Jupiter Odyssey*, D. M. Harland, Springer/Praxis 2000, p. 150

439. "Core Sizes and Internal Structure of Earth's and Jupiter's Satellites", O. L. Kuskov and V. A. Kronrod. *Icarus* 2001, 151, p. 214

440. "Implications from Galileo Observations on the Interior Structure and Chemistry of the Galilean Satellites", F. Sohl, T. Spohn, D. Breuer, and K. Nagel. *Icarus* 2002, 157, p. 106

441. "Subsurface Oceans and Deep Interiors of Medium-Sized Outer Planet Satellites and Large Trans-Neptunian Objects", H. Hussmann, F. Sohl, and T. Spohn. *Icarus* 2006, 185, p. 258

442. "Implications from Galileo Observations on the Interior Structure and Chemistry of the Galilean Satellites", F. Sohl, T. Spohn, D. Breuer, and K. Nagel. *Icarus* 2002, 157, p. 105

443. *The Giant Planet Jupiter*, J. H. Rogers, Cambridge University Press 1995, p. 331

444. "Shape, Mean Radius, Gravity Field, and Interior Structure of Callisto", J. D. Anderson, R. A. Jacobson, and T. P. McElrath. *Icarus* 2001, 153, pp. 157, 158

445. "Internal Structure of Europa and Callisto", O. L. Kuskov and V. A. Kronrod. *Icarus* 2005, 177, p. 551

446. "Implications from Galileo Observations on the Interior Structure and Chemistry of the Galilean Satellites", F. Sohl, T. Spohn, D. Breuer, and K. Nagel. *Icarus* 2002, 157, p. 106

447. "Internal Structure of Europa and Callisto", O. L. Kuskov and V. A. Kronrod. *Icarus* 2005, 177, p. 551

448. "Implications from Galileo Observations on the Interior Structure and Chemistry of the Galilean Satellites", F. Sohl, T. Spohn, D. Breuer, and K. Nagel. *Icarus* 2002, 157, p. 106

449. "Subsurface Oceans on Europa and Callisto: Constraints from Galileo Magnetometer Observations", C. Zimmer and K. K. Khurana. *Icarus* 2000, 147, p. 239

450. "Internal Structure of Europa and Callisto", O. L. Kuskov and V. A. Kronrod. *Icarus* 2005, 177, p. 550

451. "Formation of Ganymede Grooved Terrain by Sequential Extensional Episodes: Implications of Galileo Observations for Regional Stratigraphy", G. C. Collins, J. W. Head, and R, T. Pappalardo. *Icarus* 1998, 135, p. 358

452. *Jupiter Odyssey*, D. M. Harland, Springer/Praxis 2000, p. 166

453. *Jupiter Odyssey*, D. M. Harland, Springer/Praxis 2000, p. 171

454. *Jupiter Odyssey*, D. M. Harland, Springer/Praxis 2000, p. 165

455. *Jupiter Odyssey*, D. M. Harland, Springer/Praxis 2000, p. 168

456. *Jupiter Odyssey*, D. M. Harland, Springer/Praxis 2000, p. 177

457. "Galileo Views of the Geology of Callisto", R. Greeley, J. E. Klemaszewski, R. Wagner and the Galileo Imaging Team. *Planetary and Space Science* 2000, 48, p. 829

458. "Galileo Views of the Geology of Callisto", R. Greeley, J. E. Klemaszewski, R. Wagner and the Galileo Imaging Team. *Planetary and Space Science* 2000, 48, p. 829

459. "Morphology and Origin of Palimpsests on Ganymede Based on Galileo Observations", K. B. Jones, J. W. Head III, R. T. Pappalardo, and J. M. Moore. *Icarus* 2003, 164, p. 197

460. *Jupiter Odyssey*, D. M. Harland, Springer/Praxis 2000, p. 178

461. "Core Sizes and Internal Structure of Earth's and Jupiter's Satellites", O. L. Kuskov and V. A. Kronrod. *Icarus* 2001, 151, p. 214

462. "Implications from Galileo Observations on the Interior Structure and Chemistry of the Galilean Satellites", F. Sohl, T. Spohn, D. Breuer, and K. Nagel. *Icarus* 2002, 157, pp. 104, 117

463. "Shape, Mean Radius, Gravity Field, and Interior Structure of Callisto", J. D. Anderson, R. A. Jacobson, and T. P. McElrath. *Icarus* 2001, 153, p. 160

464. *Jupiter Odyssey*, D. M. Harland, Springer/Praxis 2000, p. 180

465. "Internal Structure of Europa and Callisto", O. L. Kuskov and V. A. Kronrod. *Icarus* 2005, 177, p. 563

466. "Formation of Ganymede Grooved Terrain by Sequential Extensional Episodes: Implications of Galileo Observations for Regional Stratigraphy", G. C. Collins. J. W. Head, and R. T. Pappalardo. *Icarus* 1998, 135, p. 345

467. *Jupiter: The Planet, Satellites, and Magnetosphere*, D. C. Jewett, S. Sheppard, and C. Porco. Cambridge University Press 2004, Chapter 12, p. 1

468. "An Abundant Population of Small Irregular Satellites Around Jupiter", S. S. Sheppard and D. C. Jewiit. *Nature* 2003, 423, p. 261

469. "An Abundant Population of Small Irregular Satellites Around Jupiter", S. S. Sheppard and D. C. Jewiit. *Nature* 2003, 423, p. 261

470. *Jupiter: The Planet, Satellites, and Magnetosphere*, D. C. Jewett, S. Sheppard, and C. Porco. Cambridge University Press 2004, Chapter 12, p. 6

471. "Outer Irregular Satellites of the Planets and their Relationship with Asteroids, Comets, and Kuiper Belt Objects", S. C. Sheppard. *Asteroids, Comets, Meteors: Proceedings IAU Symposium No. 229*, 2005, pp. 327–329

472. "Outer Irregular Satellites of the Planets and their Relationship with Asteroids, Comets, and Kuiper Belt Objects", S. C. Sheppard. *Asteroids, Comets, Meteors: Proceedings IAU Symposium No. 229*, 2005, p. 328

473. *Jupiter: The Planet, Satellites, and Magnetosphere*, D. C. Jewett, S. Sheppard, and C. Porco. Cambridge University Press 2004, Chapter 12, pp. 8, 9

474. *Neptune: the Planet, Rings, and Satellites*, E. D. Miner and R. R. Wessen. Springer/Praxis 2002, pp. 28, 238

475. *Neptune: the Planet, Rings, and Satellites*, E. D. Miner and R. R. Wessen. Springer/Praxis 2002, p. 174

476. "The Structure of Jupiter's Ring System as Revealed by the Galileo Imaging Experiment", M. E. Ockert-Bell, J. A. Burns, I. J. Daubar, P. C. Thomas, and J. Veverka. *Icarus* 1999, 138, p. 188

477. "The Size Distribution of Jupiter's Main Ring from Galileo Imaging and Spectroscopy", S. M. Brooks, L. W. Esposito, M. R. Showalter, and H. B. Throop. *Icarus* 2004, 170, p. 35

478. "The Size Distribution of Jupiter's Main Ring from Galileo Imaging and Spectroscopy", S. M. Brooks, L. W. Esposito, M. R. Showalter, and H. B. Throop. *Icarus* 2004, 170, pp. 43, 44

479. "The Size Distribution of Jupiter's Main Ring from Galileo Imaging and Spectroscopy", S. M. Brooks, L. W. Esposito, M. R. Showalter, and H. B. Throop. *Icarus* 2004, 170, pp. 50, 51, 52, 54, 55

480. "Galileo NIMS Near-Infrared Observations of Jupiter's Ring System", S. McMuldroch, S. H. Pilorz, G. E. Danielson, and the NIMS Science Team. *Icarus* 2000, 146, p. 2

481. "The Size Distribution of Jupiter's Main Ring from Galileo Imaging and Spectroscopy", S. M. Brooks, L. W. Esposito, M. R. Showalter, and H. B. Throop. *Icarus* 2004, 170, p. 36

482. *Jupiter Odyssey*, D. M. Harland, Springer/Praxis 2000, p. 286

483. "The Size Distribution of Jupiter's Main Ring from Galileo Imaging and Spectroscopy", S. M. Brooks, L. W. Esposito, M. R. Showalter, and H. B. Throop. *Icarus* 2004, 170, p. 36

484. "The Structure of Jupiter's Ring System as Revealed by the Galileo Imaging Experiment", M. E. Ockert-Bell, J. A. Burns, I. J. Daubar, P. C. Thomas, and J. Veverka. *Icarus* 1999, 138, p. 189

485. *Jupiter Odyssey*, D. M. Harland, Springer/Praxis 2000, p. 283

486. "The Structure of Jupiter's Ring System as Revealed by the Galileo Imaging Experiment", M. E. Ockert-Bell, J. A. Burns, I. J. Daubar, P. C. Thomas, and J. Veverka. *Icarus* 1999, 138, p. 189

487. "The Structure of Jupiter's Ring System as Revealed by the Galileo Imaging Experiment", M. E. Ockert-Bell, J. A. Burns, I. J. Daubar, P. C. Thomas, and J. Veverka. *Icarus* 1999, 138, p. 189

488. "The Structure of Jupiter's Ring System as Revealed by the Galileo Imaging Experiment", M. E. Ockert-Bell, J. A. Burns, I. J. Daubar, P. C. Thomas, and J. Veverka. *Icarus* 1999, 138, p. 188

489. "Galileo NIMS Near-Infrared Observations of Jupiter's Ring System", S. McMuldroch, S. H. Pilorz, G. E. Danielson, and the NIMS Science Team. *Icarus* 2000, 146, p. 2

490. "The Structure of Jupiter's Ring System as Revealed by the Galileo Imaging Experiment", M. E. Ockert-Bell, J. A. Burns, I. J. Daubar, Peter C. Thomas, and Joseph Veverka. *Icarus* 1999, 138, pp. 207, 208

491. "Galileo NIMS Near-Infrared Observations of Jupiter's Ring System", S. McMuldroch, S. H. Pilorz, G. Edward Danielson, and the NIMS Science Team. *Icarus* 2000, 146, p. 10

492. *Jupiter Odyssey*, D. M. Harland, Springer/Praxis 2000, pp. 284–288

493. "The Jovian Rings: New Results Derived from Cassini, Galileo, Voyager, and Earth-based Observations", H. B. Throop, C. C. Porco, R. A. West, J. A. Burns, M. R. Showalter, and P. D. Nicholson. *Icarus* 2004, 172, p. 67

494. "The Structure of Jupiter's Ring System as Revealed by the Galileo Imaging Experiment", M. E. Ockert-Bell, J. A. Burns, I. J. Daubar, P. C. Thomas, and J. Veverka. *Icarus* 1999, 138, p. 189

495. "The Size Distribution of Jupiter's Main Ring from Galileo Imaging and Spectroscopy", S. M. Brooks, L. W. Esposito, M. R. Showalter, and H. B. Throop. *Icarus* 2004, 170, p. 39

496. "The Structure of Jupiter's Ring System as Revealed by the Galileo Imaging Experiment", M. E. Ockert-Bell, J. A. Burns, I. J. Daubar, P. C. Thomas, and J. Veverka. *Icarus* 1999, 138, p. 196

497. "The Jovian Rings: New Results Derived from Cassini, Galileo, Voyager, and Earth-based Observations", H. B. Throop, C. C. Porco, R. A. West, J. A. Burns, M. R. Showalter, and P. D. Nicholson. *Icarus* 2004, 172, p. 59

498. "The Size Distribution of Jupiter's Main Ring from Galileo Imaging and Spectroscopy", S. M. Brooks, L. W. Esposito, M. R. Showalter, and H. B. Throop. *Icarus* 2004, 170, pp. 36–38, 55

499. "The Structure of Jupiter's Ring System as Revealed by the Galileo Imaging Experiment", M. E. Ockert-Bell, J. A. Burns, I. J. Daubar, P. C. Thomas, and J. Veverka. *Icarus* 1999, 138, p. 196

500. "The Jovian Rings: New Results Derived from Cassini, Galileo, Voyager, and Earth-based Observations", H. B. Throop, C. C. Porco, R. A. West, J. A. Burns, M. R. Showalter, and P. D. Nicholson. *Icarus* 2004, 172, pp. 59, 69

501. "The Structure of Jupiter's Ring System as Revealed by the Galileo Imaging Experiment", M. E. Ockert-Bell, J. A. Burns, I. J. Daubar, P. C. Thomas, and J. Veverka. *Icarus* 1999, 138, p. 196

502. "The Size Distribution of Jupiter's Main Ring from Galileo Imaging and Spectroscopy", S. M. Brooks, L. W. Esposito, M. R. Showalter, and H. B. Throop. *Icarus* 2004, 170, pp. 50–52, 54

503. "The Jovian Rings: New Results Derived from Cassini, Galileo, Voyager, and Earth-based Observations", H. B. Throop, C. C. Porco, R. A. West, J. A. Burns, M. R. Showalter, and P. D. Nicholson. *Icarus* 2004, 172, pp. 59, 60

504. "The Galileo Star Scanner Observations of Amalthea", P. D. Fieseler et al. *Icarus* 2004, 169, p. 390

505. "The Structure of Jupiter's Ring System as Revealed by the Galileo Imaging Experiment", M. E. Ockert-Bell, J. A. Burns, I. J. Daubar, P. C. Thomas, and J. Veverka. *Icarus* 1999, 138, p. 188

506. "The Structure of Jupiter's Ring System as Revealed by the Galileo Imaging Experiment", M. E. Ockert-Bell, J. A. Burns, I. J. Daubar, P. C. Thomas, and J. Veverka. *Icarus* 1999, 138, p. 203

507. "The Structure of Jupiter's Ring System as Revealed by the Galileo Imaging Experiment", M. E. Ockert-Bell, J. A. Burns, I. J. Daubar, P. C. Thomas, and J. Veverka. *Icarus* 1999, 138, pp. 206, 207

508. "The Jovian Rings: New Results Derived from Cassini, Galileo, Voyager, and Earth-based Observations", H. B. Throop, C. C. Porco, R. A. West, J. A. Burns, M. R. Showalter, and P. D. Nicholson. *Icarus* 2004, 172, pp. 70–72, 75

509. *The Giant Planet Jupiter*, J. H. Rogers, Cambridge University Press 1995, p. 402

510. *A Complete Manual of Amateur Astronomy*, C. P. Sherrod, Prentice-Hall, Inc., 1981, p. 9

511. *Visual Observations of the Planet Saturn and its Satellites: Theory and Methods*, J. L. Benton, Jr., Associates in Astronomy 1985, p. 35

512. *The Planet Jupiter*, B. M. Peek, Faber and Faber, Ltd., 2nd edition Patrick Moore 1981, p. 39

513. *Digital Astrophotography: the State of the Art*, D. Ratledge, (editor). Springer-Verlag London Limited 2005, p. 2

514. *Digital Astrophotography: the State of the Art*, David Ratledge, (editor). Springer-Verlag London Limited 2005, pp. 2, 3

515. "Processing Webcam Images with *Registax*", Cor Berrevoets. *Sky and Telescope* 2004, 107, pp. 130–165

516. *The Giant Planet Jupiter*, J. H. Rogers. Cambridge University Press 1995, p. 391
517. *The Planet Jupiter*, B. M. Peek, Faber and Faber, Ltd., 2nd edition Patrick Moore 1981, pp. 45–50
518. *The Giant Planet Jupiter*, J. H. Rogers. Cambridge University Press 1995, p. 396
519. *The Planet Jupiter*, B. M. Peek, Faber and Faber, Ltd., 2nd edition Patrick Moore 1981, p. 49
520. *The Planet Jupiter*, B. M. Peek, Faber and Faber, Ltd., 2nd edition Patrick Moore 1981, pp. 45–50
521. *The Giant Planet Jupiter*, J. H. Rogers. Cambridge University Press 1995, p. 396
522. *The Giant Planet Jupiter*, J. H. Rogers. Cambridge University Press 1995, p. 397
523. *Jupiter Observer's Handbook*, R. Schmude, Jr. The Astronomica League 2004, pp. 6–7
524. *Jupiter Observer's Handbook*, R. Schmude, Jr. The Astronomica League 2004, pp. 6–7
525. *Jupiter Observer's Handbook*, R. Schmude, Jr. The Astronomica League 2004, pp. 6–7

Index